Contents

Audio-CD cues

The following sounds can be heard on the audio-CDs that accompany this book.

DISC 1
Heart sounds

Track	Sound
1, 2	Normal S_1 and S_2 (base of heart)
3	S_1 (mitral area)
4	S_1 split so M_1 and T_1 can be heard better on expiration (tricuspid area)
5	S_1 mitral area
6	S_1 split
7	Abnormal S_1 split
8, 9	S_2 (pulmonic area)
10, 11	S_2 split during inspiration and expiration
12	Increased intensity of P_2
13	Increased intensity of A_2
14	Diminished A_2
15	S_2 split during inspiration and expiration
16	Paradoxical S_2 split heard during expiration
17	Persistent S_2 split
18	Persistent S_2 split in pulmonary hypertension
19, 20	Wide, fixed S_2 split
21	Paradoxical S_2 split
22	Fused paradoxical S_2 split
23, 24	S_3 over mitral area (S_1, S_2, S_3)
25, 26	Abnormal S_3
27, 28	Pericardial knock
29, 30	Right-sided S_3
31, 32	S_4 (S_4, S_1, S_2)

DISC 2
Breath sounds

Clinical consultants

Anne W. Davis, RN, PhD
Professor of Nursing
East Central University Department of Nursing
Ada, Okla.

Sherry L. Rogman, RN
Charge Nurse, Emergency Department
Bryan/LGH Medical Center
Lincoln, Nebr.

Kristine A. Scordo, PhD, RN,CS, ACNP
Director, Acute Care Nurse Practitioner Program
Wright State University
Dayton, Ohio

How to use this book

Auscultation Skills: Breath & Heart Sounds, Third Edition, and the accompanying audio-CDs have been designed to provide you with a practical tool to improve your ability to auscultate for these sounds and enhance your understanding of their physiology. By using this book together with the audio-CDs and referring to them time and time again, you can review each breath and heart sound until you recognize it perfectly.

This invaluable reference explains the physiology and pathophysiology of more than 40 different breath sounds and more than 50 different heart sounds. Illustrations highlight each system's anatomy as well as areas affected by — and auscultatory sites for — the sounds discussed.

The accompanying audio-CDs provide electronically generated examples of nearly all the breath and heart sounds in the book. Breath sound examples begin with normal inspiration to make it easier for you to determine where the abnormality occurs in the respiratory cycle. Heart sounds start with the first heart sound to help you better recognize abnormalities in the cardiac cycle.

We recommend that you first go through the book and audio-CDs sequentially, and then go through again to review the sounds that you find difficult. In the book, in bold type, a CD disc-and-track number appears whenever you should turn on an audio-CD to a particular track to hear a sound.

Use the handy pretests at the beginning of each chapter to check your understanding of the subject matter. If you're unsure of an answer, focus on that particular area of study as you read through the chapter. Then, after you've finished the chapter and listened to all accompanying audio-CD examples, take the posttest to help you gauge your understanding of key concepts. You'll find detailed posttest answers at the back of the book.

Throughout the text, you'll find graphic icons that call your attention to important points. *Auscultation tips* advise you of sites and techniques to detect tricky sounds. *Alerts* warn you of life-threatening auscultation findings that demand an urgent response. *Age issue* is a new feature that describes age-related differences in lung and heart anatomy and physiology as well as changes in auscultation findings across the life span.

For quick reference, we've compiled a glossary of commonly used terms and abbreviations and an easy-to-read chart of common disorders matched with their specific abnormal breath and heart sounds.

This edition also contains two new appendices. The first contains rapid-review flowcharts of seven common respiratory and cardiac findings that briefly describe what you're listening to, provide information to further direct your patient assessment, and detail appropriate diagnostic tests, treatments, and related care to give you a guidepost for further interventions.

The second new appendex is a listing of English-Spanish phrases that you'll want to consult when performing a respiratory or cardiac assessment on a patient who speaks Spanish. This resource will help you communicate more clearly with these patients and, thus, perform an accurate assessment.

ONE
Heart sound fundamentals

The heart and auscultation

1

PRETEST ▶▶▶

1. Where is the heart located in the chest, and what are the heart's dimensions in the average adult?
2. List the chambers of the heart.
3. Name the atrioventricular and semilunar heart valves.
4. Describe the path of blood flow through the heart.
5. What are systole and diastole?
6. How do electrical impulses stimulate the heart muscle to contract?
7. Describe the path of the heart's conduction system.
8. Name the precordial sites recommended for auscultation of heart sounds.
9. Which qualities should you look for in choosing a stethoscope?
10. Which sound frequency is transmitted best by the bell of the stethoscope? By the diaphragm of the stethoscope?
11. What should you do to improve your ability to hear heart sounds more clearly during auscultation?
12. Which maneuvers enhance heart sounds during auscultation?

Anatomy and physiology

HEART LOCATION AND SIZE

The heart is a powerful muscle. Its primary functions are to pump deoxygenated blood to the lungs, where carbon dioxide-oxygen (CO_2-O_2) exchange takes place, and to pump oxygenated blood throughout the body. The top, or superior, part of the heart is called the *base*; the inferior, lateral part of the heart is called the *apex*. The average-sized adult heart is about 4¾" (12 cm) long from the base to the apex and about 3⅛" (8 cm) wide at its widest point. The

Dimensions of average-sized adult heart

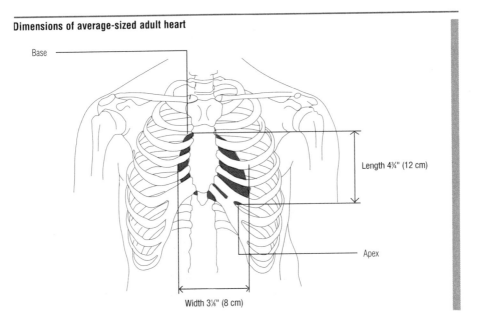

Base

Length 4¾" (12 cm)

Apex

Width 3⅛" (8 cm)

heart gradually increases in size from infancy through early adulthood and usually is slightly larger in males than in females.

RIGHT AND LEFT SIDES

Anatomically, the heart is divided into two functional sides: the right side and the left side, which are separated by the septum. Each side is further divided into chambers. The right atrium and right ventricle are the right-sided chambers; the left atrium and left ventricle, the left-sided chambers.

Each chamber has unidirectional valves that separate the chambers and control blood flow. The right atrioventricular (AV), or *tricuspid,* valve separates the right atrium and right ventricle; the left AV, or *mitral,* valve separates the left atrium and left ventricle. The semilunar pulmonic valve separates the right ventricle and pulmonary artery, and the semilunar aortic valve separates the left ventricle and aorta.

BLOOD FLOW

Deoxygenated, or *venous,* blood from nearly every tissue of the body flows into the right atrium via the superior and inferior venae cavae. Venous blood from the heart tissue drains via the coronary sinus directly into the right atrium. The deoxygenated blood collects there until the tricuspid

Internal structures of the heart

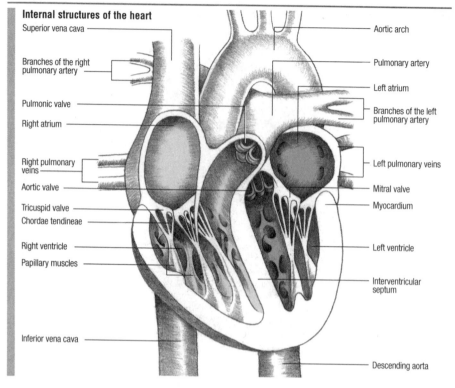

Superior vena cava

Branches of the right pulmonary artery

Pulmonic valve

Right atrium

Right pulmonary veins

Aortic valve

Tricuspid valve

Chordae tendineae

Right ventricle

Papillary muscles

Inferior vena cava

Aortic arch

Pulmonary artery

Left atrium

Branches of the left pulmonary artery

Left pulmonary veins

Mitral valve

Myocardium

Left ventricle

Interventricular septum

Descending aorta

valve opens, and then it flows into the right ventricle. From the ventricle, it flows through the open pulmonic valve into the pulmonary artery, which distributes the deoxygenated blood to both lungs.

The complex process of CO_2-O_2 exchange occurs in the alveoli. Then the oxygenated blood flows back to the heart via the pulmonary veins and collects in the left atrium. From there it flows through the opened mitral valve into the left ventricle. When the aortic valve opens, the oxygenated blood flows into the aorta, which transports it to all body tissues.

HEART AS A PUMP

A healthy heart in a 154-lb (70-kg) adult pumps about 1⅝ gal (6 L) of blood per minute. Because the total blood volume in a healthy adult is about 1⅜ gal (5.2 L), the heart pumps the total blood volume of the body in less than a minute. It performs this task continuously throughout an individual's life.

Pathway of deoxygenated blood through right heart to lungs

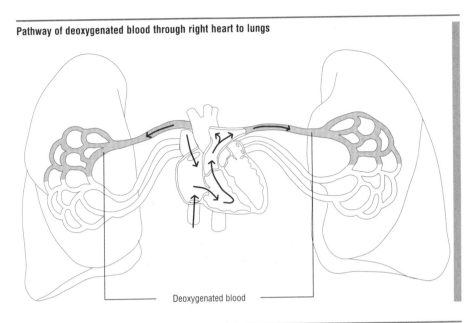

Deoxygenated blood

Pathway of oxygenated blood from lungs through left heart

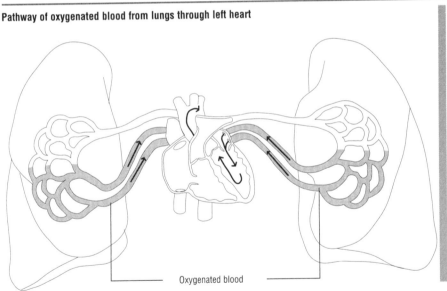

Oxygenated blood

An active process

The flow of deoxygenated and oxygenated blood through the heart, lungs, and rest of the body is an active process. The heart must generate enough pressure to pump blood through the arterial circulation. It does this by rhythmically contracting (systole) and relaxing (diastole).

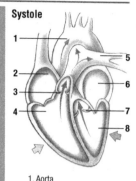

Systole

1. Aorta
2. Right atrium
3. Pulmonic valve
4. Right ventricle
5. Pulmonary artery
6. Left atrium
7. Aortic valve
8. Left ventricle

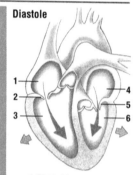

Diastole

1. Right atrium
2. Tricuspid valve
3. Right ventricle
4. Left atrium
5. Mitral valve
6. Left ventricle

During systole, heart muscle contraction raises the intracardiac pressures enough to systematically open the semilunar valves and pump a certain volume of blood. During diastole, the heart muscle relaxes, allowing the chambers to fill with blood again. This entire sequence is called *the cardiac cycle.*

The amount of blood pumped into the aorta each minute (cardiac output) depends largely on stroke volume. Stroke volume is the amount of blood pumped during each ventricular contraction.

AGE ISSUE *As a person without heart disease ages, his heart usually becomes slightly larger and loses its contractile strength and efficiency. By age 70, cardiac output at rest has usually diminished by 30% to 35%.*

CONDUCTION SYSTEM

Electrical impulses are primarily responsible for the rhythmic pumping action of the heart. These impulses are discharged automatically from specialized cardiac cells that stimulate the heart muscle to contract. The specialized cardiac cells and fibers within the conduction system work together to produce a contraction. The impulses fired from the sinoatrial (SA) node, located just below the entrance of the superior vena cava in the right atrium, normally initiate the cardiac cycle, and they usually travel along specific pathways.

After the impulse

When the electrical impulse leaves the SA node, it travels along the atrial conduction pathways located in the atrial walls, initiating atrial systole (or atrial contraction). The impulse rapidly reaches the AV node, located near the AV junction. From the AV node, the impulse travels to the bundle of His located in the interventricular septum. Here, the bundle of His divides into the right and left bundle branches. These branches further divide into the tiny Purkinje fibers located throughout the ventricular walls. As the impulse travels through the bundle branches and Purkinje fibers, it initiates ventricular systole, or ventricular contraction.

AGE ISSUE *As the myocardium of the aging heart becomes more irritable, extrasystoles may occur, along with sinus arrhythmias and bradycardias. In addition, increased fibrous tissue infiltrates the SA node and internodal tracts, possibly causing atrial fibrillation and flutter.*

Cardiac conduction

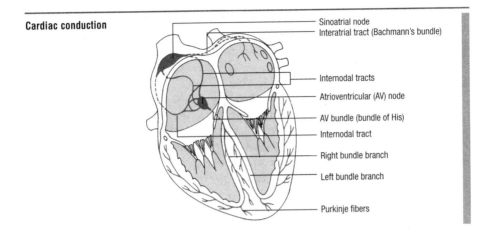

Sinoatrial node
Interatrial tract (Bachmann's bundle)
Internodal tracts
Atrioventricular (AV) node
AV bundle (bundle of His)
Internodal tract
Right bundle branch
Left bundle branch
Purkinje fibers

CARDIAC CYCLE

Several mechanical events occurring simultaneously during the cardiac cycle enable the heart to pump oxygenated blood throughout the body.

During atrial diastole, when the right atrium and left atrium are relaxed, blood flows into the atria. When the pressure in the atria exceeds the pressure in the ventricles, the tricuspid and mitral valves open, allowing blood to flow rapidly into the relaxed ventricles. The SA node, the pacemaker in a normally functioning heart, initiates an electrical impulse. The impulse is conducted via internodal pathways to both atria and results in atrial contraction. This contraction—atrial systole—pushes more blood from the atria to complete ventricular filling. The impulse is slowly conducted through the AV node, which allows time for the atria to contract completely and fill the ventricles with blood. From the AV node, the impulse is rapidly conducted through the bundle of His and its branches. The Purkinje fibers receive the impulse and directly stimulate the ventricular myocardium. A forceful contraction of both ventricles opens the pulmonic and aortic valves and ejects blood into the pulmonary artery and aorta.

Unidirectional flow

The sequential and synchronized contraction (systole) and relaxation (diastole) of the atria and ventricles, along with the opening and closing of competent, properly functioning valves, ensure a unidirectional flow of blood through the heart and allow the heart to effectively move blood to all areas of the body quickly and repeatedly. However, this isn't a silent process. Auscultating heart sounds can clue you in to events in the cardiac cycle, such as the opening and closing of heart valves.

Traditional auscultatory areas

Auscultatory areas

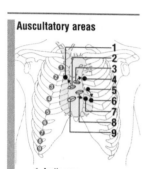

1. Aortic area
2. Aortic valve
3. Pulmonic valve
4. Pulmonic area
5. Erb's point
6. Mitral area
7. Tricuspid area
8. Mitral valve
9. Tricuspid valve

The heart's base normally lies beneath the sternum at the third intercostal space. Its apex is usually located between the fourth and sixth intercostal spaces, near or medial to the left midclavicular line. The apex is where apical pulses are taken and is usually the point of maximum impulse (PMI), the site where the heartbeat is seen or felt most strongly. However, to auscultate for heart sounds and murmurs, specific precordial sites are used. These sites are the aortic, pulmonic, tricuspid, and mitral areas and Erb's point. Each area is named after the heart valve primarily responsible for the sounds that are heard best in that location.

 AGE ISSUE *An infant's heart is positioned more horizontally in the chest cavity than an adult's heart. As a result, the apex is at the fourth left intercostal space and, until age 4, the apical impulse is to the left of the midclavicular line. By age 7, the heart is in the same position as the adult heart.*

 AGE ISSUE *In the elderly patient, degenerative bony prominences can shift cardiac anatomy downward or laterally.*

LOCATING THE PRECORDIAL SITES

The aortic area is located at the second right intercostal space, close to the sternal border. This area is where sounds of the aortic valve and aorta are heard best.

The pulmonic area is located at the second left intercostal space, close to the sternal border. This area is where sounds of the pulmonic valve and pulmonary artery are heard best.

The tricuspid area is located at the fifth left intercostal space, close to the sternal border. This area is where sounds of the tricuspid valve and right ventricle are heard best.

The mitral area is located at or near the fifth intercostal space, just medial to the left midclavicular line. This area, directly over the left ventricle, is sometimes referred to as the apical area, or apex. The mitral area is where sounds of the mitral valve and left ventricle are heard best.

Erb's point differs from the four other areas in that it isn't named after a heart valve, but it's a good location to hear sounds of aortic and pulmonic origin. This area is located at the third left intercostal space, close to the sternal border.

Alternative auscultatory areas

Although the traditional areas are a good reference for auscultating heart sounds, it's important to know that the areas described can overlap extensively. Sounds that the heart valves produce can be heard throughout the precordium. Therefore, alternative auscultatory areas should also be considered when listening to the heart. (See *Alternative auscultatory areas, page 10.*)

LOCATING THE ALTERNATIVE AUSCULTATORY AREAS

The left ventricular area is located from the second to the fifth intercostal spaces and extends from the left sternal border to the left midclavicular line. When the left ventricle is enlarged, this area extends in all directions. When the right ventricle is enlarged, this area is displaced to the left. The aortic component of the second heart sound A_2 and a left ventricular S_3 and S_4 are heard here. Murmurs of mitral or aortic stenosis and mitral or aortic regurgitation, functional middiastolic rumble, and murmurs associated with hypertrophic obstructive cardiomyopathy (formerly known as *idiopathic hypertrophic subaortic stenosis*) are best heard here.

The right ventricular area is located from the second to the fifth intercostal spaces centered over the sternum. In severe right ventricular enlargement, this area may extend as far lateral as the point of maximum impulse. A right ventricular S_3 or S_4 and the opening snap of the tricuspid valve are heard here. Murmurs of tricuspid stenosis or regurgitation and ventricular septal defect are also heard here.

The left atrial area is located from the second to the fourth intercostal space at the left sternal border. The murmur of mitral regurgitation is best heard here.

The right atrial area is located from the third to the fifth intercostal spaces at the right sternal border and may extend to or beyond the right midclavicular line in severe right atrial enlargement. This area is where the murmur of tricuspid regurgitation is heard best.

Alternative auscultatory areas

Aortic area
Pulmonic area
Left atrial area
Left ventricular area

Right atrial area
Right ventricular area

1
2
3
4
5
6
7
8
9

The aortic area is located from the right second inter-costal space to the apex of the heart. The aortic ejection click and the aortic component of the second heart sound A_2 are heard best in this area. Murmurs of aortic stenosis and regurgitation, increased aortic flow, dilation of the ascending aorta, and abnormalities of the carotid and subcla-vian arteries are also heard here.

The pulmonic area is located at the second and third intercostal spaces close to the left of the sternum, but it may be higher or lower. The pulmonary component of the second heart sound, P_2, and the pulmonary ejection sound are heard here as well as the murmurs of pulmonary stenosis and regurgitation, increased pulmonary flow, patent ductus arteriosus, and stenosis of the main branch of the pulmonary artery.

Auscultatory techniques

THE STETHOSCOPE

The stethoscope allows you to hear heart vibrations that are transmitted to the chest wall. These vibrations have various characteristics: Some are easily heard; others are less distinct. Consequently, the better the stethoscope, the better your chances of hearing subtle differences in the sounds.

Choose your stethoscope carefully. The chest piece should have a diaphragm and a bell. The diaphragm is de-

Parts of the stethoscope

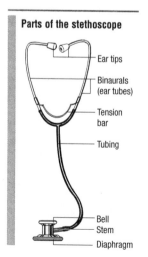

Ear tips

Binaurals
(ear tubes)

Tension
bar

Tubing

Bell
Stem
Diaphragm

signed to transmit high-pitched sounds more clearly, and the bell is designed to transmit low-pitched sounds more clearly. A flat, adult-sized diaphragm should be about 1⅜" (3.5 cm) in diameter and should be smooth, thin, and stiff enough to filter out low-frequency sounds. The bell should be about 1" (2.5 cm) in diameter and deep enough so that it doesn't fill with tissue when placed on the chest wall. The tubing length can vary from 10" to 15" (25 to 38 cm), and its inside diameter should be ⅛" (3 mm) — or ³⁄₁₆" (5 mm) if the tubing is longer. The earpieces should fit inside the ears comfortably (covering the external ear canal), and the stethoscope shouldn't be too heavy because its weight can interfere with your ability to hear sounds accurately.

AGE ISSUE *Be sure to use the appropriate size stethoscope when auscultating the heart sounds of a pediatric patient. Most stethoscopes have an adapter to attach a smaller diaphragm. The bell is useful for pediatric and thin patients.*

USING THE STETHOSCOPE

The way you use the stethoscope's diaphragm and bell affects the quality of the heart sounds and murmurs you hear during auscultation. To maximize the effectiveness of auscultation, you must hold the diaphragm and bell correctly.

To use the diaphragm, firmly grasp the metal area between the bell and the diaphragm with your finger and thumb, and press down firmly on the chest wall. You may also place your fingertips on the rim of the bell, and press down firmly against the chest wall. Apply enough pressure so that an indentation remains on the skin after you remove the stethoscope.

The bell, however, should touch the chest wall only lightly. If you exert too much pressure when listening with the bell, the stretched skin beneath it will act as a diaphragm, filtering out low-pitched sounds. To hold the bell correctly, grasp the diaphragm's outer edges with the index finger and thumb, and gently rest the bell on the chest wall. Look at the skin around the edges of the bell to see if too much pressure is being applied. If any signs of indentation appear on the skin, relax the downward pressure.

Also, remember to hold the chest piece so that the tubing isn't touching the patient or you; such contact creates sounds that interfere with effective listening.

Auscultatory sequence

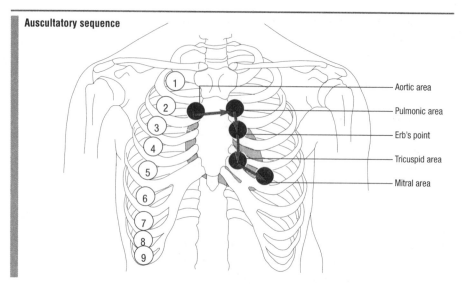

Aortic area
Pulmonic area
Erb's point
Tricuspid area
Mitral area

AUSCULTATORY SEQUENCE

During auscultation, listen with the stethoscope over all five auscultatory areas, first with the diaphragm and then with the bell. Be sure to listen through several cardiac cycles to allow yourself enough time to become oriented to the sounds and to hear any subtle changes. Try to establish a pattern, and follow it methodically each time you perform auscultation.

Performing the assessment

Before starting auscultation, place the patient in a supine position. You may elevate the head of the bed slightly if it's more comfortable for the patient. Then stand to the patient's right. Listen first with the diaphragm over the aortic area. Try to recognize the characteristic *lub-dub* sound of each cardiac cycle. After listening through several cycles, inch the diaphragm toward the pulmonic area. Try to move the diaphragm without losing track of the *lub-dub* sound. Then listen over the pulmonic area, Erb's point, and the tricuspid area. Continue in this manner until you reach the mitral area.

To locate the PMI, turn the patient slightly to the left, placing him in a partial left lateral recumbent position. This position brings the apex of the heart closer to the chest wall. Closely observe the chest wall for any sign of the heartbeat, which may appear as a rhythmic bulging. Then use your fingertip pads to palpate for the heartbeat between the fourth and the sixth intercostal spaces, near the left midclavicular line. The chest wall site where the heartbeat is seen and felt is the PMI. In a healthy adult, the PMI and the mitral area are usually located at the same site.

After listening over the PMI with the diaphragm of the stethoscope, turn the chest piece so that you can use the bell. Then, listen over the PMI with the bell. Reverse the auscultatory sequence, inching the bell over all the auscultatory areas.

If the patient can tolerate it, always auscultate for heart sounds in supine, seated, and standing positions. Thus, after auscultating with both the bell and the diaphragm while the patient is in a supine position, have him sit up and lean forward slightly. Then repeat the entire auscultatory sequence with the diaphragm and the bell.

Enhancement techniques

ENSURING THE PROPER ENVIRONMENT

A quiet, comfortable environment is recommended during auscultation for heart sounds. Privacy and a warm room are beneficial because the patient's entire chest is exposed throughout the examination. Shivering or any muscular movement interferes with heart sound transmission, making it impossible to accurately detect subtle characteristics of the sounds. Close the door and try to eliminate any mechanical noises in the room.

SOME HELPFUL MANEUVERS

Several maneuvers can enhance the heart sounds heard during auscultation. Besides position changes, you can also use maneuvers that increase or decrease blood flow through the heart. For example, you can have the patient squat or stand, hold his breath, or raise his legs while in a supine position. Another common technique is to have the patient cough several times or perform Valsalva's maneuver. During this maneuver, the patient attempts to exhale forcefully against a closed glottis.

Another method that enhances the sounds heard during auscultation is to have the patient increase the depth of breathing. This decreases the respiratory rate and makes it easier to distinguish sounds originating in either the right or left side of the heart. For example, during inspiration, venous return is enhanced, accentuating right-sided cardiac events; during expiration, left-sided blood flow is enhanced, accentuating left-sided cardiac events.

You can help the patient by raising your hand slowly when you want him to inhale and then lowering your hand slowly when you want him to exhale. Caution the patient to breathe smoothly and continuously. If he holds his breath, he may produce a Valsalva-type maneuver, which will negate the cardiac responses being assessed.

◀◀◀ POSTTEST

1. What is the heart's primary function?
2. Where are the base and the apex of the heart located?
3. How is the heart divided anatomically?
4. Name the four unidirectional valves.
5. Describe blood flow through the heart.
6. Give the approximate total blood volume of a healthy adult and the time it takes the heart to pump this total blood volume.
7. What happens in the heart during systole and diastole?
8. What are stroke volume and cardiac output?
9. Describe the conduction of electrical impulses in the heart.
10. Briefly describe the heart's pumping action.
11. What are the names and locations of the five auscultatory areas used during auscultation for heart sounds?
12. Which characteristics should you look for when choosing a stethoscope?
13. Describe how to use the diaphragm and bell of the stethoscope correctly.
14. Describe the auscultatory sequence.
15. How do you locate the point of maximum impulse?
16. What can you do to enhance heart sounds?

Heart sound dynamics

2

PRETEST ▶▶▶

1. How do heart sounds originate?
2. Name the two basic heart sounds and the valves associated with each.
3. How many characteristics are used to describe heart sounds?
4. How does the electrocardiogram (ECG) correlate with the heart's electrical activity?
5. What is a phonocardiogram?
6. How can you accurately document the heart sounds heard during auscultation?

Heart sound origins

The heart sounds heard through a stethoscope during auscultation are generated by vibrations from the heart's walls and valves and from turbulent blood flow. Normally, the heart's walls and valves are compliant as they move in response to pressure and volume changes during the cardiac cycle.

BLOOD FLOW
At the beginning of ventricular diastole, the ventricles are relaxed, and the aortic valve closes slightly before the pulmonic valve. For a brief time, both the atrioventricular (AV) and semilunar valves are closed. During this period, called *isovolumic relaxation*, intraventricular pressures continue to fall. However, the opening of the tricuspid and mitral valves follows quickly, and blood begins to flow passively from the atria into the ventricles. As the ventricles become distended, the rate of ventricular filling decreases. Then, as the atria

contract (from stimulation of the atrial myocardium by an electrical impulse from the sinoatrial [SA] node), ventricular volume is slightly boosted. The resulting increase in ventricular pressures causes the mitral and tricuspid valves to begin closing.

During atrial systole, the electrical impulse moves more slowly through the AV node and into the bundle of His. The impulse proceeds rapidly through the bundle branches and the Purkinje fibers, activating ventricular systole. Early in ventricular systole, intraventricular pressures increase above intra-atrial pressures, and the mitral and tricuspid valves close suddenly. Normally, the mitral valve closes slightly before the tricuspid valve.

For another brief period, both the AV and semilunar valves are closed. During this period, known as *isovolumic contraction,* ventricular pressures increase, becoming higher than the pressures in the pulmonary artery and aorta. The aortic and pulmonic valves open, and blood is ejected through them. Normally, the aortic valve opens slightly before the pulmonic valve.

After a period of maximum ejection of blood from the ventricles, the ventricles begin to relax. As intraventricular pressures fall rapidly, a brief backflow of blood from the pulmonary artery and aorta toward the ventricles occurs. Reduced intraventricular pressures and the temporary backflow of blood are associated with closure of the aortic and pulmonic valves at the end of systole.

Basic heart sounds

A normally functioning heart produces two basic heart sounds: S_1 and S_2. (♦1-1) S_1, the first heart sound, is heard at the beginning of systole. It's associated with closure of the mitral and tricuspid valves and the increasing pressures within the ventricles that cause the moving valve leaflets and cord structures to decelerate. S_1 is generated by the closing of AV valves and the vibrations associated with tensing of the chordae tendineae and the ventricular walls.

At the end of ventricular systole, ventricular pressures fall rapidly, causing a slight backflow of blood from the aorta and pulmonary artery. The decrease in ventricular pressures to a level below that of aortic and pulmonic systolic pressures and the temporary backflow of blood and recoil events are associated with closure of the aortic and pul-

Valves producing first heart sound (S₁)

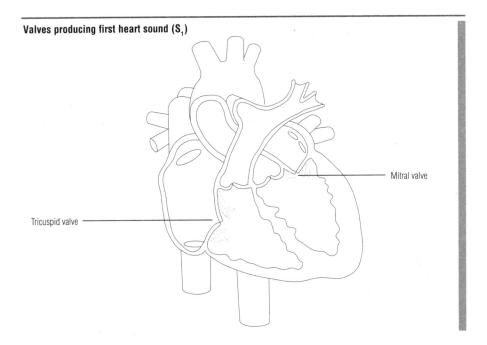

Mitral valve

Tricuspid valve

monic valves. (See *Valves producing second heart sound [S₂]*, page 18.) The vibrations associated with these events produce the second heart sound, S_2, which marks the end of ventricular systole. (◆**1-2**)

CHARACTERISTICS
Every heart sound has six different characteristics that need to be assessed during auscultation. These are *location, intensity, duration, pitch, quality,* and *timing.* If you keep these characteristics in mind each time you perform auscultation, you'll ensure a complete assessment of heart sounds, and you'll be able to provide the information necessary for accurate and complete documentation. Also, because these terms are used universally, all health care professionals will be able to understand your auscultatory findings.

Location
A sound's location is the anatomic area on the patient's chest wall where the sound is heard best. Bony structures and landmarks, such as the right and left midclavicular lines, are used to describe the precise location. For example, you might document that S_1 was heard best over the mitral area.

Valves producing second heart sound (S$_2$)

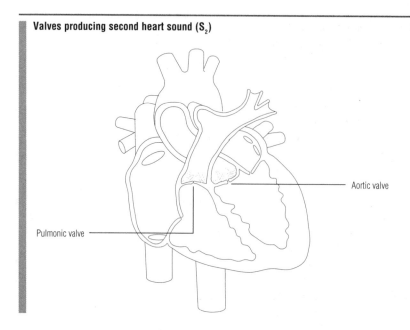

Aortic valve

Pulmonic valve

Intensity

Intensity refers to the sound's loudness during auscultation. Usually, a sound's intensity is determined somewhat subjectively, based on experience. However, when abnormalities are present, intensity can be determined electronically by recording a phonocardiogram (PCG) and by measuring the amplitude of the sound's vibrations. A PCG is a graphic representation of heart sounds that can be compared with the sounds heard during auscultation and with the heart's electrical activity recorded on the patient's ECG.

Keep in mind that heart sound intensity is related to the pressures generated and the blood flow velocity within the heart. It's also related to the patient's size, body build, and chest configuration. For example, a slender patient has a thinner chest wall; therefore, sounds are transmitted more easily through a slender patient's chest wall than through an obese patient's chest wall. Sounds can also be diminished by certain media, such as pericardial fluid or emphysematous lung tissue.

Duration

Sound duration refers to the length of time the sound is heard; it can be described as either short or long. Remember, heart sounds are brief vibrations that mark the beginning and end of systole. A sound's duration affects whether you hear it as a click, a snap, or a murmur. For example, murmurs are longer vibrations usually associated with blood flow during systole or diastole.

Pitch
The pitch of a sound is determined by the frequency of its vibrations. High-frequency sounds, like the notes of a piccolo, are best heard with the diaphragm of the stethoscope; low-frequency sounds, like the notes of a tuba, are best heard with the bell of the stethoscope.

Quality
Sound quality is determined by the combination of its frequencies. It may be described as sharp, dull, booming, snapping, blowing, harsh, or musical.

Timing
The timing of a sound refers to when it's heard during the cardiac cycle — that is, during systole or diastole.

Putting them together
By using these six characteristics, you can describe precisely every heart sound, whether normal or abnormal. You can document what you hear, where you hear it, how you hear it, and when you hear it. Complete documentation also provides other health care professionals with invaluable information that helps them recognize subtle changes in the patient's heart sounds.

Depolarization and repolarization

The sequential and rhythmic process of depolarization and repolarization is referred to as the heart's electrical activity. Electrical activity occurs at the cellular level in every contractile cell of the myocardium.

Initially, the contractile cell is polarized (in a resting state). Depolarization begins when the cell membrane becomes permeable to the flow of sodium ions into the cell. Repolarization begins when calcium ions move into the cell and potassium ions begin to move out of the cell. Then, the accumulated intracellular sodium and calcium ions are extruded while the lost potassium is restored to the cell by the sodium-potassium pump, completing repolarization. The

The ECG

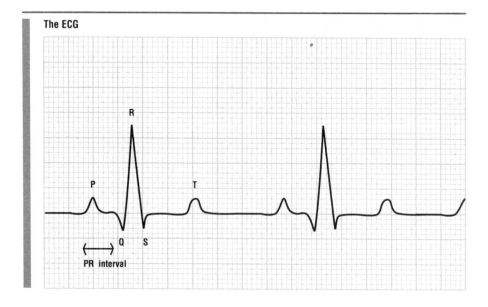

cell is polarized to its original ionic state, and the myocardium relaxes.

THE ECG

Heart sounds are produced by mechanical events that occur in response to an electrical impulse originating in the SA node. This electrical impulse travels through the myocardium, activating the atria and ventricles. The ECG documents the timing and amplitude of the heart's electrical activity but doesn't reflect mechanical events in the heart.

The ECG correlates with the heart's electrical activity in the following manner. The SA node fires, and the impulse spreads through the atria. The P wave, which is part of the ECG waveform, represents atrial depolarization. Atrial contraction is stimulated by and closely follows atrial depolarization. The QRS complex represents electrical depolarization of the ventricles. Atrial repolarization isn't seen in the ECG because it's hidden in the PR segment and the QRS complex. The T wave represents ventricular repolarization.

The impulse travels through the AV node, the bundle of His, the bundle branches, and the Purkinje fibers before the ventricles contract. The time between atrial depolarization and ventricular depolarization is recorded in the ECG as the

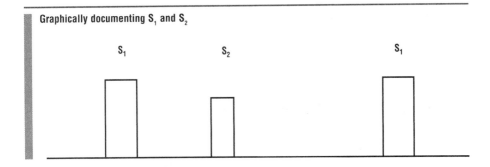

Graphically documenting S₁ and S₂

S_1 S_2 S_1

PR interval: It begins at the onset of the P wave and lasts until the onset of the QRS complex. The time interval of the PR interval correlates with the time interval between atrial contraction and ventricular contraction.

ECG'S RELATIONSHIP TO FIRST AND SECOND HEART SOUNDS

The first and second heart sounds directly correlate with a patient's ECG. S_1 normally occurs just after the QRS complex; S_2, at the end of the T wave. (See *Correlation of S_1 and S_2 with ECG*, page 22.)

Documenting heart sounds

The characteristics of each heart sound must be thoroughly documented. A PCG provides one method of graphically representing heart sounds. (See *PCG and ECG showing S_1 and S_2*, page 22.) However, it isn't used routinely for every patient; therefore, a more practical graphic method of documenting heart sounds is recommended.

The intensity, duration, and timing of heart sounds can be represented graphically as rectangular blocks placed perpendicularly on a horizontal baseline. The block's height corresponds to the heart sound's intensity (the higher the intensity, the higher the block); the block's width, to the duration of the sound.

The proximity of one block on the baseline to the next block represents the time interval between one sound and another. The spacing between the blocks can also be used to specify the relationship of a particular heart sound to systole or diastole.

Correlation of S₁ and S₂ with ECG

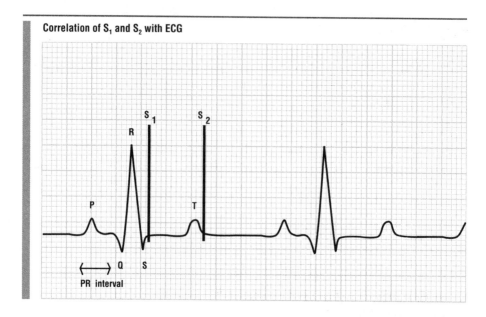

PCG and ECG showing S₁ and S₂

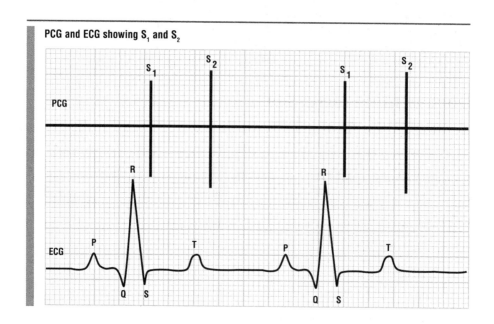

◀◀◀ POSTTEST

1. How are heart sounds generated?
2. Describe what happens in the heart from the beginning of ventricular diastole to the beginning of atrial systole.
3. Describe what happens in the heart from the beginning of atrial systole to the beginning of ventricular systole.
4. Describe what happens in the heart from the beginning of ventricular systole to the beginning of ventricular diastole.
5. How is S_1 generated?
6. How is S_2 generated?
7. What is a heart sound's location?
8. What is a heart sound's intensity?
9. What is a heart sound's duration?
10. What is a heart sound's pitch?
11. What is a heart sound's quality?
12. What is a heart sound's timing?
13. Briefly explain the process of depolarization and repolarization.
14. Describe how the ECG correlates with the heart's electrical activity.
15. When do S_1 and S_2 occur in relation to the ECG waveform?
16. Describe a graphic method used to document heart sounds.

TWO
Normal heart sounds

The first heart sound

3

PRETEST ▶▶▶

1. What produces the first heart sound (S₁)?
2. What are the two basic components of S₁?
3. Where is S₁ heard best?
4. Describe S₁, using six characteristics.
5. What is a widened S₁ split?

Normal S_1

Valves involved in S₁ sound production

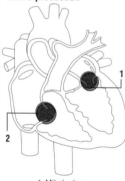

1. Mitral valve
2. Tricuspid valve

The cardiac vibrations associated with closure of the mitral and tricuspid valves produce the first heart sound, S_1. Normally, only two components of the first heart sound are audible. The first component, referred to as M_1, is associated with closure of the mitral valve; the second component, T_1, is associated with closure of the tricuspid valve. Both valves close at the beginning of ventricular systole, but the mitral valve usually closes slightly ahead of the tricuspid valve.

AUSCULTATORY AREA AND RELATIONSHIP TO ECG

M_1 and T_1 are usually perceived as a single sound called S_1, which is heard best near the apex of the heart over the mitral area, using the diaphragm of the stethoscope. A single sound is heard because left-sided heart sounds are normally more intense at this site. S_1 occurs just after the QRS complex on the electrocardiogram (ECG) waveform. **(◆1-3)**

As you inch the stethoscope from the mitral area toward the tricuspid area, without losing track of S_1, the M_1 and T_1 components of S_1 may become evident. Expiration makes them easier to hear. **(◆1-4)** T_1 trails M_1 slightly and is softer; it's heard best near the left sternal border. The timing of

PCG and ECG showing S₁

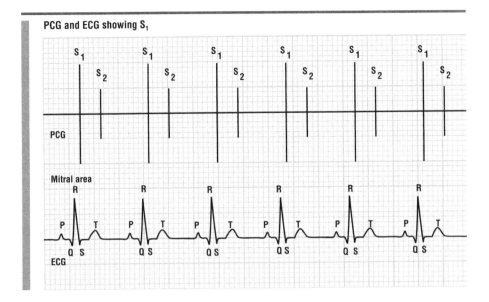

M₁ and T₁ with the QRS complex remains the same. In many patients, only the first heart sound is heard because M₁ and T₁ are separated by a 20-millisecond — or shorter — pause, which the human ear perceives as one sound. (See *PCG and ECG showing normal S₁ split,* page 28.)

AUSCULTATION TIP *If S₁ is difficult to identify, palpate for the carotid pulse while performing auscultation. S₁ will occur just before you feel the carotid pulse.*

Sound characteristics
S₁ is usually heard best near the heart's apex over the mitral area. Its intensity directly relates to the force of ventricular contraction and the ECG PR interval. The shorter the PR interval, the more widely open the mitral and tricuspid leaflets are at the onset of ventricular contraction. Thus, the distance they must be moved is greater, causing more intense vibrations when they close and a louder S₁. S₁ is short in duration. You can hear its high pitch best with the diaphragm of the stethoscope. The quality of S₁ is somewhat dull. S₁ timing coincides with the beginning of ventricular systole and just precedes a palpable carotid pulse. **(◆1-5)** The characteristics of the normally split S₁ into M₁ and T₁ are the same. **(◆1-6)**

Auscultatory area for S₁

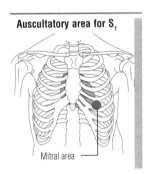

Mitral area

Auscultatory area for M₁ and T₁

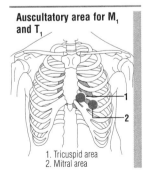

1. Tricuspid area
2. Mitral area

PCG and ECG showing normal S₁ split

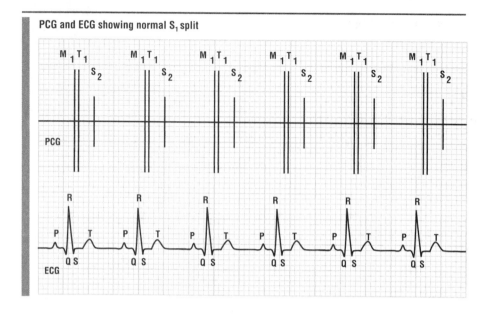

PCG and ECG showing abnormal S₁ split

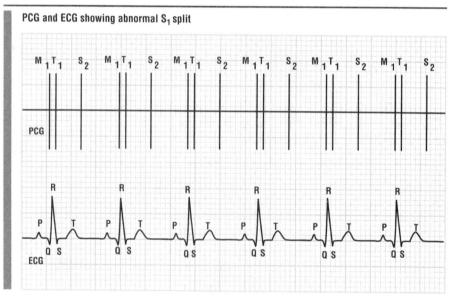

Enhancing S₁

S_1 can be enhanced by sympathetic stimulation, such as that provided by a brief period of exercise.

Abnormal S_1 split

WIDENED S_1 SPLIT

The normal M_1-T_1 split heard over the tricuspid area widens when electrical activation and contraction of the right ventricle are delayed. Such a delay causes delayed tricuspid valve closure. (**◆1-7**) This abnormal finding is associated with complete right bundle-branch block, left ventricular ectopic beats, epicardial pacing of the left ventricle, tricuspid stenosis, atrial septal defect, Ebstein's anomaly, and left ventricular tachycardia.

Auscultatory area for abnormal S_1 split

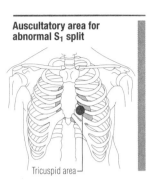

Tricuspid area

◆◀◀ POSTTEST

1. What produces S_1?
2. What are the components of S_1?
3. Where is S_1 heard best during auscultation, and how does it change as the stethoscope is inched from one area to another?
4. Describe the characteristics of S_1.
5. How can S_1 be enhanced?
6. When does an abnormal S_1 split occur?
7. Which conditions are associated with an abnormal S_1 split?

The second heart sound

4

PRETEST ▶▶▶

1. How does the second heart sound (S_2) differ from the first heart sound (S_1)?
2. What are the components of S_2?
3. Describe the characteristics of S_2.
4. What is an S_2 split?
5. Which types of S_2 splits may be heard during auscultation?

Normal S_2

Valves involved in S_2 sound production

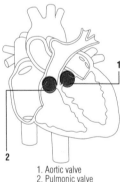

1. Aortic valve
2. Pulmonic valve

The cardiac vibrations associated with closure of the aortic and pulmonic valves produce the second heart sound, S_2. S_2 is usually louder than S_1 at the heart's base and usually slightly higher in pitch than S_1 at the heart's apex.

S_2, like S_1, has two basic components: the aortic (A_2) component and the pulmonic (P_2) component. Both valves close at the end of ventricular systole. Normally, the aortic valve closes slightly ahead of the pulmonic valve.

Closing pressure is higher in the aorta than in the pulmonary artery; therefore, A_2 usually occurs earlier and is louder than P_2. (◆1-8) Furthermore, A_2 can be heard over the entire precordium, whereas P_2 is usually auscultated over the pulmonic area.

AUSCULTATORY AREA AND RELATIONSHIP TO ECG
To hear both components of S_2, listen carefully with the diaphragm of the stethoscope over the pulmonic area. Listen for S_2 just after the T wave in the patient's electrocardiogram (ECG).

PCG and ECG showing S$_2$

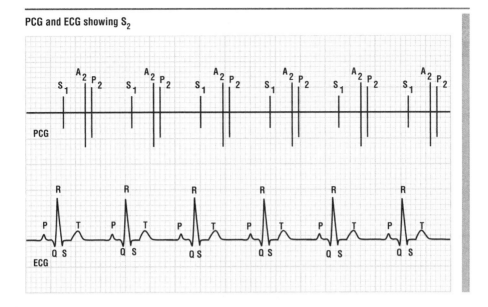

Sound characteristics

S$_2$ is usually heard best near the heart's base, over the pulmonic area or over Erb's point. Its intensity directly relates to the amount of closing pressure in the aorta and pulmonary artery. S$_2$ is slightly shorter in duration than S$_1$. You can hear its high pitch best with the diaphragm of the stethoscope. The quality of S$_2$ is somewhat booming; its timing coincides with the end of ventricular systole. (◆1-9)

Auscultatory area for S$_2$

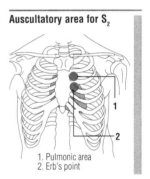

1. Pulmonic area
2. Erb's point

S$_2$ split sounds

NORMAL S$_2$ SPLIT

The splitting of S$_2$ into the A$_2$ and P$_2$ components is heard best during inspiration over Erb's point.

AGE ISSUE *Normal splitting of S$_2$ in children may be heard in the second left intercostal space. The splitting of S$_2$ in elderly persons may be more difficult to hear because of widening of the anteroposterior chest diameter.*

Remember, inspiration causes an increase in venous return to the right side of the heart. This increased venous return prolongs right ventricular ejection time. Inspiration also reduces pressure in the pulmonary artery, resulting in a delayed P$_2$ or closure of the pulmonic valve. A decrease in blood flow to the left side of the heart occurs simultaneously, resulting in a shorter left ventricular ejection time.

Auscultatory area for normal S$_2$ split

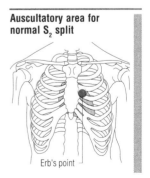

Erb's point

PCG and ECG showing respiratory changes in normal S₂ split

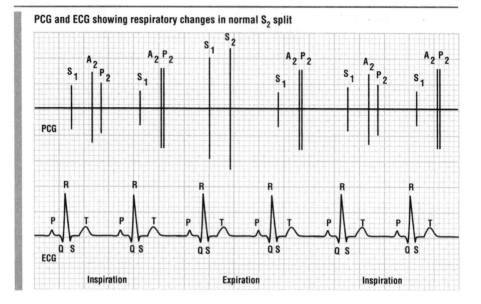

Therefore, A_2 is heard earlier than P_2. (**◆1-10**) This means that a normal S_2 split is heard during inspiration, and the A_2 and P_2 components fuse during expiration.

While listening to S_2, you must determine the intensity of the A_2 and P_2 components and note the duration of the A_2-P_2 interval as well as its relationship to the respiratory cycle. (**◆1-11**) You should also document the S_2 split, noting whether the A_2-P_2 interval increases or decreases during inspiration and expiration.

Changes in A_2 and P_2 intensity

The intensity of A_2 and P_2 changes proportionally with the difference in pressure gradients across the closed aortic and pulmonic valves. For example, P_2 may be louder than normal in conditions associated with elevated pulmonary artery diastolic pressure, as occurs in some patients with heart failure, mitral stenosis, Eisenmenger's syndrome, or other congenital heart diseases. When P_2 increases in intensity, it's sometimes heard over the mitral area and along the left sternal border. (**◆1-12**)

A_2 intensity increases when diastolic pressure in the aorta increases. (**◆1-13**) This commonly happens during exercise; during states of excitement such as extreme fear; in hyperkinetic conditions, such as thyrotoxicosis, fever, and pregnancy; and in systemic hypertension. However, if the patient has left ventricular decompensation, ventricular

PCG and ECG showing abnormal S₂ split with no respiratory variation in a patient with elevated pulmonary artery diastolic pressure

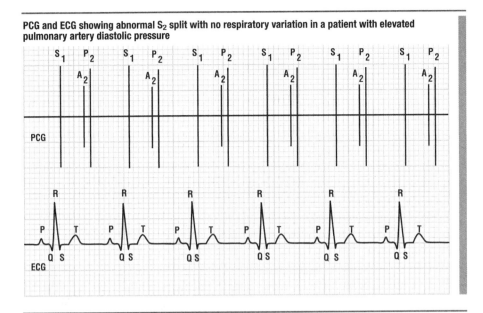

PCG and ECG showing abnormal S₂ splits during respiratory cycle

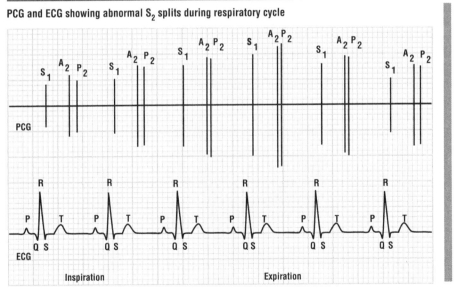

relaxation is slower, and the pressure gradients may not be great enough to produce an accentuated A_2.

Conversely, A_2 intensity may be diminished in conditions that alter the development of diastolic pressure gradients, such as aortic regurgitation and hypotension. (◆1-14) A_2 intensity also decreases when ventricular dysfunction is present, such as after an acute myocardial infarction. In this condition, P_2 may become louder if pulmonary artery pressure rises and systemic pressure falls. A_2 is also softer or absent when aortic valve motion is restricted, as in severe aortic stenosis.

ABNORMAL S_2 SPLITS

Abnormal splitting is related to valvular dysfunction, to alterations in blood flow to or from the ventricles, or to both. These changes may cause the normal S_2 split to be absent during both phases of the respiratory cycle. Thus, only a single S_2 is heard over Erb's point. In another case, the split sounds may persist through inspiration and expiration with little or no respiratory variation. (◆1-15) The split sounds may also be heard paradoxically on expiration. The A_2-P_2 intervals vary, as does the intensity of A_2 and P_2 during the respiratory cycle. (◆1-16) Changes in S_2 splits are usually most noticeable at the beginning of inspiration and expiration. (See *PCG and ECG showing abnormal S_2 split with no respiratory variation in a patient with elevated pulmonary artery diastolic pressure* and *PCG and ECG showing abnormal S_2 split during respiratory cycle,* page 33.)

Absent S_2 split

The P_2 component may not be heard during auscultation over Erb's point in the patient with severe pulmonic stenosis. Consequently, S_2 remains a single sound during both inspiration and expiration. A normal S_2 split may also be absent if the A_2 sound masks the P_2 sound or vice versa — for example, when one sound is significantly louder than the other, making splitting inaudible. This phenomenon occurs in patients with pulmonary hypertension. On the other hand, systemic hypertension causes A_2 to be delayed and to fuse with P_2 during inspiration. In patients with an increased anteroposterior chest dimension, the P_2 intensity may be so diminished that only A_2 is audible.

Persistent S_2 split

A persistent S_2 split occurs when A_2 and P_2 don't fuse into one sound during expiration, even though some respiratory variation in the intensity of A_2 and P_2 is heard. (◆1-17) This persistent A_2-P_2 splitting during expiration usually results from early aortic valve closure or delayed pulmonic valve closure.

PCG and ECG showing persistent S₂ split

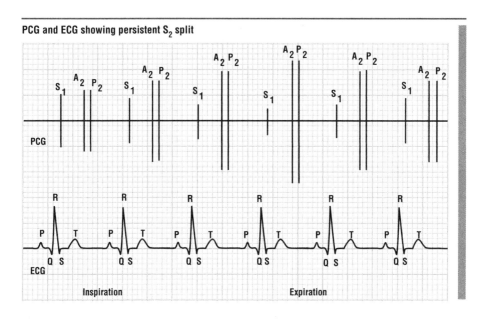

PCG and ECG showing narrowed S₂ split in a patient with chronic pulmonary hypertension

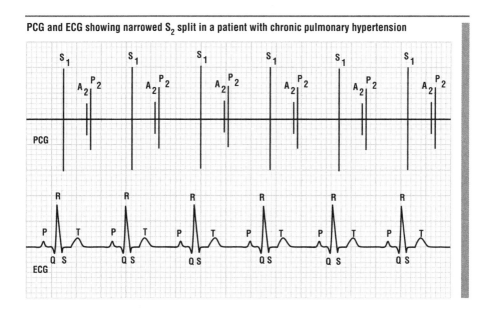

Early aortic valve closure is associated with shortened left ventricular systole, which occurs in patients with mitral regurgitation, ventricular septal defects, or cardiac tamponade. Delayed pulmonic valve closure occurs when right ventricular systole is prolonged in patients with chronic

PCG and ECG showing wide, fixed S₂ split

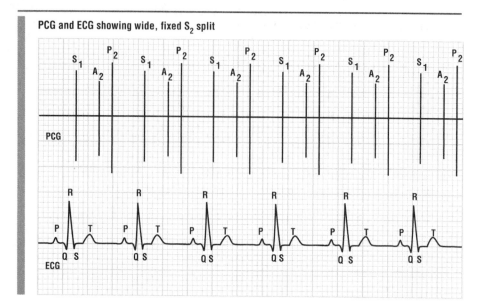

pulmonary hypertension. In these patients, the A_2-P_2 split is heard, but the interval between the A_2 and P_2 components is narrowed. (See *PCG and ECG showing narrowed S_2 split in a patient with chronic pulmonary hypertension,* page 35.) **(◆1-18)**

Delayed electrical activation of the right ventricle
Another cause of persistent A_2-P_2 splitting throughout expiration is delayed electrical activation of the right ventricle, which will delay P_2. This phenomenon is commonly found in patients with right bundle-branch block, left ventricular epicardial pacing, or left ventricular ectopic beats.

Wide, fixed S₂ split
A prolonged right ventricular ejection time produces expiratory A_2-P_2 splits that are widened and that persist even when the patient is seated. **(◆1-19)** This occurs in patients with atrial septal defects, acute pulmonary hypertension secondary to massive pulmonary emboli, or pulmonic stenosis. The P_2 component may not be audible in patients with severe pulmonic stenosis.

A wide, fixed S_2 split is also associated with the pulmonary vascular bed's increased capacitance (ability to receive blood volume and the decreased resistance that accompanies it). **(◆1-20)** This phenomenon occurs in patients with idiopathic dilation of the pulmonary artery or with

PCG and ECG showing paradoxical S₂ split during expiration

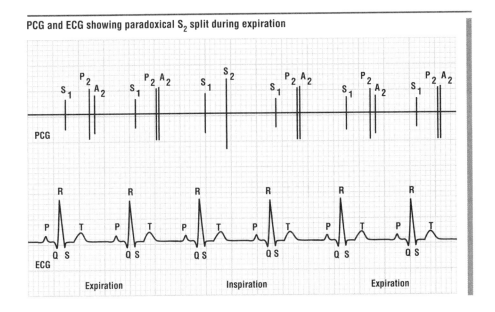

atrial septal defects. This split does not change with respiration.

Paradoxical S₂ split

In a paradoxical, or reversed, S₂ split, P₂ precedes A₂, and the split sounds are heard during expiration instead of inspiration. (♦1-21) Delayed aortic valve closure almost always causes this phenomenon. If A₂ is delayed during expiration, it may follow P₂, causing an S₂ split; if A₂ is delayed during inspiration, A₂ and P₂ fuse because inspiration normally delays P₂. Therefore, S₂ is heard as a single sound during inspiration instead of a normally split sound. (♦1-22)

Delayed A₂ is commonly seen in patients with delayed activation of the left ventricle caused by left bundle-branch block. It's also associated with right ventricular ectopic beats and with right ventricular endocardial pacing. Prolonged left ventricular systole can also cause a delay in A₂. The same paradoxical P₂-A₂ split may be heard in patients with left-sided heart failure, patent ductus arteriosus, coarctation of the aorta, tricuspid regurgitation, ischemic heart disease, or early right ventricular activation across accessory tracks, as in Wolff-Parkinson-White syndrome.

Left ventricular pressure overload, which occurs in patients with systemic hypertension or hypertrophic cardiomyopathy, may also cause paradoxical S_2 split. Likewise, aortic stenosis may lead to paradoxical splitting of A_2 and P_2; however, if the stenosis is severe, the A_2 component may be inaudible.

With left ventricular volume overload, which is commonly associated with aortic regurgitation or patent ductus arteriosus, A_2 may be delayed, and a paradoxical S_2 split may be heard; however, this is rather uncommon.

◀◀◀ POSTTEST

1. What produces S_2?
2. What are the two basic components of S_2?
3. Where is S_2 heard best during auscultation?
4. Describe the characteristics of S_2.
5. How does respiration affect S_2?
6. Which clinical conditions change the intensity of A_2 and P_2?
7. Describe an abnormal S_2 split.
8. How do severe pulmonic stenosis and pulmonary hypertension affect the S_2 split?
9. Describe a persistent S_2 split.
10. Describe a wide, fixed S_2 split.
11. Describe a paradoxical S_2 split.

The third and fourth heart sounds

5

PRETEST ▶▶▶

1. What are left ventricular diastolic filling sounds?
2. How do the third and fourth heart sounds (S_3 and S_4) differ from the first and second heart sounds (S_1 and S_2)?
3. Describe a normal S_3.
4. Describe an abnormal S_3.
5. What is a right-sided S_3?
6. What is a ventricular gallop, or gallop rhythm?
7. What is an atrial diastolic gallop?
8. What is a pericardial knock?
9. Describe a normal S_4.
10. Describe an abnormal S_4, a summation gallop, and a right-sided S_4.

Ventricular filling sounds

The first and second heart sounds, S_1 and S_2, mark the beginning and end, respectively, of ventricular systole. In healthy individuals, these two sounds and their components are relatively easy to hear during auscultation and are best heard with the diaphragm of the stethoscope.

The two left ventricular diastolic filling sounds, S_3 and S_4, are sometimes heard over the mitral area. These sounds differ from S_1 and S_2 in that they're low-frequency sounds produced by ventricular filling, rather than associated with valve closure.

Third heart sound

Area producing S₃

1. Mitral valve
2. Left ventricle
3. Tricuspid valve

NORMAL S₃

Occasionally, a physiologic third heart sound, S_3, is heard during auscultation. This heart sound, considered normal in healthy individuals younger than age 20 and in athletic young adults, is caused by vibrations occurring during rapid, passive ventricular filling.

AGE ISSUE *In patients older than age 40, an S_3 is likely the result of heart failure or ventricular volume overload caused by valvular disease.*

Early in diastole, after isovolumic relaxation, the mitral and tricuspid valves open and the ventricles fill and expand. In children and young adults, the left ventricle is normally compliant, permitting rapid filling. The left ventricle responds to this rapid filling with an abrupt change in wall motion that causes a sudden decrease in blood flow. These events generate vibrations that are responsible for the physiologic S_3. The more vigorously the left ventricle expands, the greater the likelihood that an S_3 will occur.

A physiologic S_3 is also commonly audible in patients with high-output conditions, in which rapid ventricular expansion, caused by increased blood volume, is present. Anemia, fever, pregnancy, and thyrotoxicosis are conditions that cause rapid ventricular expansion, resulting in an S_3. An S_3 is also commonly heard in young, slender individuals during periods of excessive catecholamine release.

Auscultatory area and relationship to ECG

Auscultatory area for S₃

Mitral area

S_3 is typically heard best with the bell of the stethoscope over the mitral area, and it can usually be palpated over the same area. It's heard best during expiration when blood flow into the left ventricle is increased. (◆**1-23**) S_3 is heard early in diastole and follows S_2 by 0.14 to 0.20 second. The relationship of S_3 to the electrocardiogram (ECG) waveform can be seen easily on a phonocardiogram (PCG); in the ECG, S_3 occurs during the TP interval just after the T wave. Normally, the S_2-S_3 interval is reliably constant.

Sound characteristics

S_3 is usually heard best near the apex of the heart, over the mitral area. In some patients, it's soft and faint in intensity and difficult to hear; in others, it may be loud and easy to hear. S_3 has a short duration, and it may occur only intermittently during every third or fourth heartbeat.

PCG and ECG showing physiologic S₃

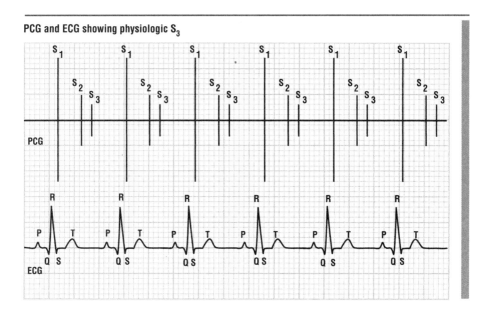

S₃ has a low pitch that's heard best with the bell of the stethoscope, and it usually has a dull, thudlike quality. Its timing is closely related to S₂, which is heard just after the T wave; S₃ follows S₂ by less than 0.20 second. **(◆1-24)**

Enhancing S₃

Because S₃ is associated with blood volume and velocity, it can be intensified by maneuvers that increase stroke volume, such as elevating the patient's legs from a recumbent position or having the patient exercise briefly or cough several times. A physiologic S₃ usually disappears with maneuvers that decrease venous return, such as having the patient sit up or stand.

AUSCULTATION TIP *To enhance your ability to auscultate and palpate for an S₃, place the patient in a partial left lateral recumbent position.*

ABNORMAL S₃

The S₁, S₂, S₃ sequence is referred to as a *ventricular gallop* or *gallop rhythm.* **(◆1-25)** An abnormal S₃ has the same sound characteristics, is heard over the same location, and has the same timing in relation to S₂ as a physiologic S₃. The differences between the two are related to the patient's age, clinical condition, or both. Also, an S₃ gallop rhythm usually persists despite maneuvers that decrease venous return. **(◆1-26)**

Area producing abnormal S₃

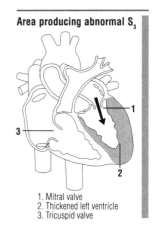

1. Mitral valve
2. Thickened left ventricle
3. Tricuspid valve

PCG and ECG showing abormal S₃ in a patient older than age 20 with a clinical condition

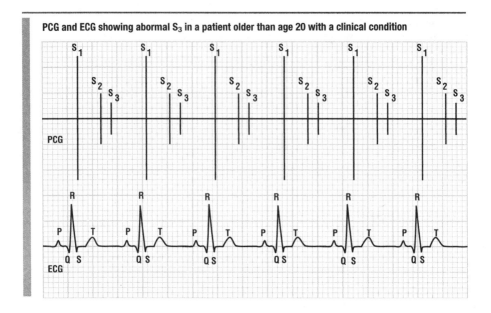

PCG and ECG showing pericardial knock

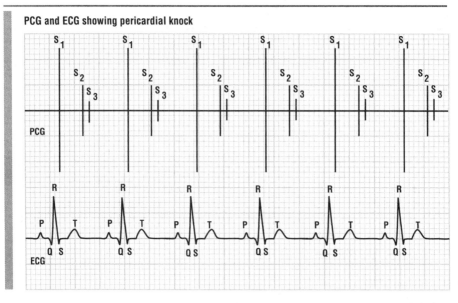

An abnormal S_3 is heard in conditions associated with increased blood volume and increased inflow velocity into the left ventricle. It's heard with or without an increase in ventricular diastolic pressure. Consequently, patients with mitral regurgitation and heart failure have an abnormal S_3.

This heart sound can also be heard during increased blood flow through the mitral valve, which occurs in patients with ventricular septal defects, patent ductus arteriosus, or severe aortic regurgitation.

✿ **AGE ISSUE** *The left ventricle becomes less compliant with age. Decreased ventricular compliance leads to pressure changes, which cause slowed left ventricular isovolumic relaxation and reduced blood flow velocity into the ventricle. Thus, an audible S_3 in a patient older than age 20 is considered abnormal unless the individual is a highly trained athlete.*

An abnormal S_3 in patients with constrictive pericarditis is called a *pericardial knock.* This type of abnormal S_3 occurs closer to S_2. The interval between S_2 and S_3 is usually less than 0.14 second. (◆1-27) The pericardial knock is easier to hear than a physiologic or abnormal S_3 and is heard best over the third, fourth, and fifth intercostal spaces along the left sternal border using the diaphragm of the stethoscope. (◆1-28) Inspiration usually intensifies the pericardial knock.

◎ **AUSCULTATION TIP** *To increase the intensity of an abnormal S_3, have the patient perform maneuvers that increase the venous return to the heart, such as raising the legs while in a recumbent position.*

Right-sided S_3

Because the right ventricle is normally much more compliant than the left ventricle, its filling shouldn't cause the vibrations that would create a right-sided S_3. However, in some patients, an S_3 originates in the right ventricle instead of the left. When it does, it's always considered abnormal. (◆1-29)

A right-sided S_3 is heard best over the third, fourth, and fifth intercostal spaces along the left sternal border and over the epigastric area. It's more prominent during inspiration because of increased blood flow into the right ventricle. (See *PCG and ECG showing right-sided S_3,* page 44.)

Many patients with an enlarged right ventricle have a right-sided S_3. This heart sound is audible in patients with right-sided heart failure, pulmonic regurgitation, or severe tricuspid insufficiency. (◆1-30)

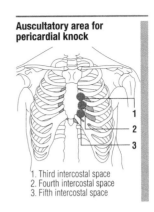

Auscultatory area for pericardial knock

1. Third intercostal space
2. Fourth intercostal space
3. Fifth intercostal space

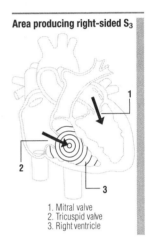

Area producing right-sided S_3

1. Mitral valve
2. Tricuspid valve
3. Right ventricle

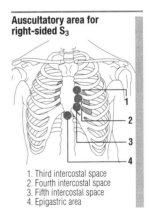

Auscultatory area for right-sided S_3

1. Third intercostal space
2. Fourth intercostal space
3. Fifth intercostal space
4. Epigastric area

PCG and ECG showing right-sided S₃

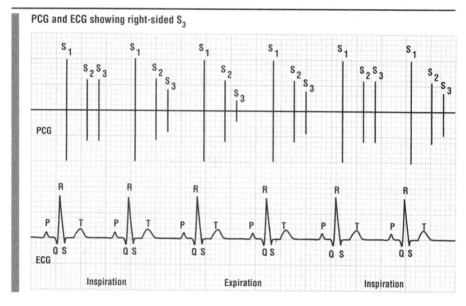

PCG and ECG showing S₄

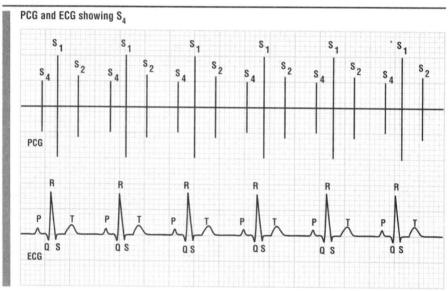

Fourth heart sound

NORMAL S₄

By the end of diastole, the ventricles are nearly full; atrial contraction further stretches and fills the ventricles. The vibrations caused by this filling and stretching in late diastole generate an additional heart sound, S₄, sometimes called an *atrial diastolic gallop.* (**◆1-31**) An S_4 is usually abnormal except in highly trained young athletes with physiologic left ventricular hypertrophy.

Because S_4 is associated with atrial contraction, it isn't produced in conditions in which atrial systole doesn't occur, such as atrial fibrillation.

Auscultatory area and relationship to ECG
S_4 is usually heard best with the bell of the stethoscope during expiration near the heart's apex over the mitral area; it also may be palpable in this area. The relationship of S_4 to S_1 can be seen easily on the ECG waveform: S_4 occurs during the PR interval and precedes S_1.

Sound characteristics
S_4 is usually located near the heart's apex over the mitral area; occasionally, it's also palpable over this area. In some patients, it's faint in intensity and difficult to hear, but in others it's loud and easy to hear. S_4 is relatively short in duration and may occur only intermittently, during every third or fourth heartbeat. It has a low pitch and a thudlike quality that's heard best with the bell of the stethoscope. Its timing is presystolic: S_4 precedes S_1 and occurs during the PR interval. (**◆1-32**)

AUSCULTATION TIP *To enhance S_4, place the patient in the partial left lateral recumbent position, which brings the heart closer to the chest wall.*

ABNORMAL S_4
An abnormal S_4 is usually associated with increased mean left atrial pressure caused by a noncompliant left ventricle. This auscultatory finding is heard in patients with hypertension (the most common cause), hypertrophic cardiomyopathy, cardiomyopathies, or ischemic heart disease; during an episode of angina; and during or after an acute myocardial infarction. When an S_4 is heard in a patient with hypertension, the systolic blood pressure usually exceeds 160 mm Hg or the diastolic pressure exceeds 100 mm Hg.

An abnormal S_4 can also accompany certain volume overload conditions, such as hyperthyroidism, certain anemias, and sudden severe mitral regurgitation.

AGE ISSUE *In children, an S_4 may accompany congenital abnormalities associated with heart failure, such as a large ventricular septal defect and patent ductus arteriosus. It may also result from sickle cell anemia.*

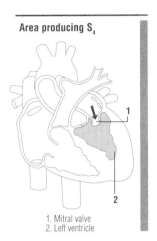

Area producing S_4

1. Mitral valve
2. Left ventricle

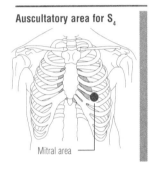

Auscultatory area for S_4

Mitral area

Area producing right-sided S₄

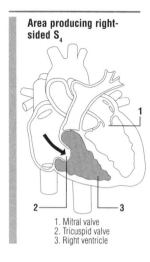

1. Mitral valve
2. Tricuspid valve
3. Right ventricle

Auscultatory area for right-sided S₄

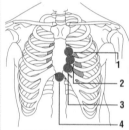

1. Third intercostal space
2. Fourth intercostal space
3. Fifth intercostal space
4. Epigastric area

Summation gallop

Normally, S_4 precedes S_1 by an appreciable interval that correlates with the PR interval on the ECG. However, in patients with first-degree atrioventricular block, the P wave occurs early in diastole, and S_4 may occur during the early rapid diastolic filling period. Likewise, in tachycardia, S_4 may be superimposed on S_3 during early rapid filling. If either condition exists, the S_4 fuses with S_3 to become a single diastolic filling sound called a *summation gallop,* which may be louder than S_4, S_3, or S_1. (**◆1-33**)

Right-sided S₄

An S_4 generated in the right ventricle is called a *right-sided S_4.* (**◆1-34**) It's commonly heard in patients with conditions that increase pressure in the right ventricle by more than 100 mm Hg, such as pulmonic stenosis or pulmonic hypertension. This heart sound is heard best over the third, fourth, and fifth intercostal spaces along the left sternal border and over the epigastric area with the patient in a supine position; it's more audible during inspiration.

DIFFERENTIATING S₄ FROM AN S₁ SPLIT

Distinguishing an S_1 split (M_1-T_1) from an S_4-S_1 sequence is sometimes difficult; however, here are some ways to help you do so. An S_1 split is heard best between the mitral and tricuspid areas with the diaphragm of the stethoscope, whereas an S_4 is heard best over the mitral area with the bell of the stethoscope and usually isn't audible with the diaphragm. Also, an S_4 may be palpable.

AUSCULTATION TIP *To increase the intensity of S_4, have the patient perform maneuvers that increase left atrial pressure, such as the handgrip exercise.*

S₃ and S₄: Similarities and differences

S_3, like S_4, is low pitched and may be faint or loud and heard only intermittently. Both sounds are heard best over the mitral area using the bell of the stethoscope, with the patient in a partial left lateral recumbent position. Both sounds are intensified by expiration and can be enhanced by maneuvers that increase stroke volume, such as elevating the legs while in a recumbent position or performing handgrip exercises.

PCG and ECG showing summation gallop

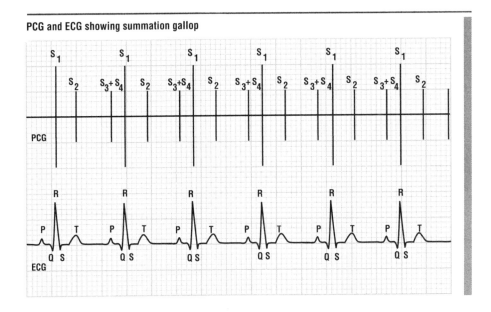

PCG and ECG showing right-sided S₄

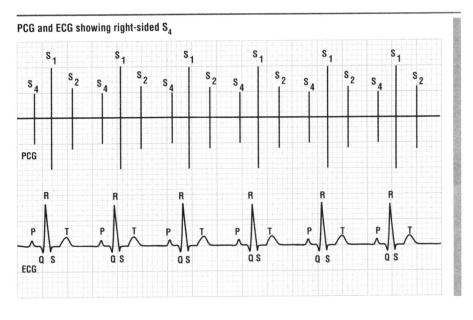

◀◀◀ POSTTEST

1. What are the two left ventricular diastolic filling sounds, and how do they differ from S_1 and S_2?
2. How is an S_3 produced?
3. List the conditions associated with an S_3.
4. Where and when is S_3 heard best?
5. Describe the characteristics of S_3.
6. Which maneuvers can be used to enhance S_3?
7. An abnormal S_3 has the same sound characteristics, is heard at the same location, and has the same timing in relation to S_2 as a physiologic S_3. What is the difference between the two sounds?
8. Which conditions are associated with an abnormal S_3?
9. Describe the characteristics of a right-sided S_3.
10. How is an S_4 produced?
11. Describe the characteristics of S_4.
12. How can an S_4 be enhanced?
13. In which conditions is an abnormal S_4 heard?
14. Describe a summation gallop.
15. Describe a right-sided S_4.
16. How can you differentiate between an S_4 and an S_1 split?
17. Compare and contrast S_3 and S_4.

THREE
Abnormal heart sounds

Other diastolic and systolic sounds

6

PRETEST ▶▶▶

1. What causes an opening snap?
2. How can you differentiate an opening snap from a second or a third heart sound (S_2 or S_3)?
3. What causes systolic ejection sounds?
4. What is the difference between a pulmonic ejection sound and an aortic ejection sound?
5. What is a midsystolic click (MSC)?
6. How can you differentiate an MSC from other heart sounds?

Opening snaps

At the end of ventricular systole, the aortic and pulmonic valves close, generating the second heart sound, S_2. S_2 is followed by a brief period of isovolumic relaxation; during this time ventricular pressures fall. When ventricular pressures are less than atrial pressures, the mitral and tricuspid valves open. Normally, the opening of these heart valves is inaudible. However, if the mitral valve leaflets become stenotic or abnormally narrowed while remaining somewhat mobile, they create an *opening snap* (OS). (◆1-35) The OS sound is generated by the maximum, but limited, opening of the stenotic leaflets and usually indicates the beginning of the diastolic murmur associated with mitral stenosis. If the mitral valve becomes severely calcified and inflexible, the OS disappears.

AUSCULTATORY AREA AND RELATIONSHIP TO ECG

An OS is heard best with the diaphragm of the stethoscope near the heart's apex over the mitral area or just medial to it.

Area producing opening snap

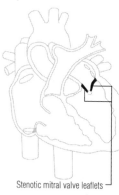

Stenotic mitral valve leaflets

PCG and ECG showing opening snap

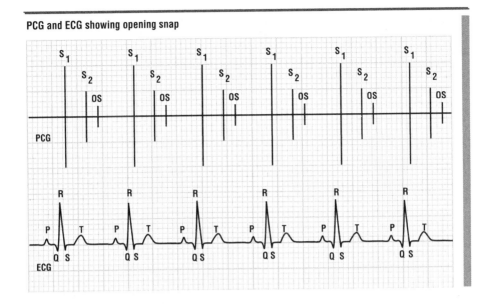

It's transmitted widely across the precordium and can typically be heard over the aortic, pulmonic, and tricuspid areas as well. An OS is higher in pitch than S_2; it may be as loud as or louder than S_2.

An OS occurs early in ventricular diastole; thus, it's heard just after the T wave in the electrocardiogram (ECG) waveform.

Sound characteristics
The intensity of an OS varies among patients, and it's usually easy to hear during auscultation. An OS has a short duration. It has a high pitch that's heard best with the diaphragm of the stethoscope, and a sharp, snaplike quality. Its timing is closely related to S_2; an OS occurs early in ventricular diastole, just after the stenotic mitral valve opens. (♦1-36)

DIFFERENTIATING AN OS FROM S_2
An OS occurs early in diastole and consequently may be confused with P_2 or S_3. One characteristic of an OS that helps distinguish it from P_2 is its timing: The A_2-P_2 interval is normally shorter than the A_2-OS interval. Also, when the patient stands, the A_2-P_2 interval narrows, whereas the A_2-OS interval widens.

Another characteristic is that the A_2-OS interval remains constant throughout respiration, whereas the A_2-P_2

Auscultatory area for opening snap

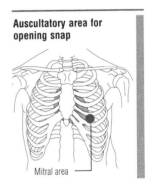

Mitral area

interval normally widens during inspiration and narrows during expiration. During inspiration, three distinct sounds can usually be heard over the pulmonic area; therefore, the sequence must be A_2, P_2, OS. In contrast, during expiration, the A_2-P_2 interval narrows or fuses, forming one sound. This creates an S_2-OS interval.

Finally, P_2 isn't usually heard over the mitral area. Therefore, if you hear a split S_2 over this area, it may be an S_2 and an OS.

DIFFERENTIATING AN OS FROM S_3

Distinguishing an OS from an S_3 is difficult in some patients, especially in those with mild mitral stenosis when the A_2-OS interval is wider than usual and the OS is somewhat softer. One characteristic of an OS that helps distinguish it from an S_3 is its timing: The A_2-S_3 interval is usually longer than the A_2-OS interval. Also, S_3 is a low-frequency sound that's heard best over the mitral area using the bell of the stethoscope. In contrast, an OS produces a high-frequency sound that's more widely transmitted across the precordium and heard best using the diaphragm of the stethoscope.

Another characteristic of an OS is that its intensity usually isn't affected by having the patient stand, whereas S_2 intensity can be decreased by standing. If the OS is affected, it's intensified. Finally, if the murmur typically heard in patients with mitral stenosis is present, you can confirm that the sound is an OS.

AUSCULTATION TIP *S_3 is usually louder during expiration than during inspiration and is typically palpable; an OS doesn't vary in intensity with respiration.*

Area producing ejection sound

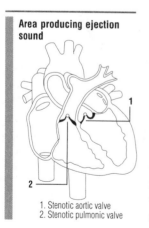

1. Stenotic aortic valve
2. Stenotic pulmonic valve

Systolic ejection sounds

Just as an OS is caused by stenotic mitral valve leaflets, a systolic ejection sound (ES) is caused by the opening of a stenotic aortic or pulmonic valve.

A systolic ES usually occurs early in systole after S_1 and isovolumic contraction. It's commonly associated with ventricular ejection and the maximum opening of a stenotic, yet mobile, aortic or pulmonic valve. If the valve is severely stenotic because of calcification, a systolic ES — like an OS — won't be produced.

A systolic ES is considered an abnormal condition, regardless of whether it originates on the heart's right or left side. It may also be caused by sudden distention of an already dilated aorta or pulmonary artery and by forceful ventricular ejection from pulmonary or systemic hypertension.

AUSCULTATORY AREA AND RELATIONSHIP TO ECG

A systolic ES is usually a brief, high-frequency sound that's heard best using the diaphragm of the stethoscope. It may be heard near the heart's base over the aortic or pulmonic area, over Erb's point, or near the apex over the mitral area. A systolic ES occurs just after the QRS complex on the ECG waveform.

PULMONIC EJECTION SOUNDS

A pulmonic ejection sound (PES), heard best over the pulmonic area, is the only right-sided heart sound that increases in intensity during expiration and diminishes or disappears during inspiration. (**◆1-37**) In a normal heart, inspiration increases right ventricular volume, causing the pulmonic valve to dome toward the pulmonary artery, which decreases the sound's intensity. During expiration, right ventricular volume is decreased, the valve is less domed, and its opening produces a louder snap. In a patient with pulmonary artery dilation or pulmonary hypertension, a PES may not vary in intensity during respiration.

Sound characteristics
A PES is usually heard best near the base of the heart over the pulmonic area. Its intensity is soft but may be equal to or greater than that of S_1; it has a short duration. It has a high pitch heard best with the diaphragm of the stethoscope and a sharp, or clicklike, quality. Its timing is closely related to S_1. A PES occurs early in ventricular systole, just after the opening of a stenotic pulmonic valve (**◆1-38**), and is heard just after the QRS complex. (See *PCG and ECG showing pulmonic ejection sound*, page 54.)

AUSCULTATION TIP *To differentiate between a PES and an S_1, remember that a PES is heard in the pulmonic area and varies with respiration, whereas a split S_1 is heard in the tricuspid area and doesn't vary with respiration.*

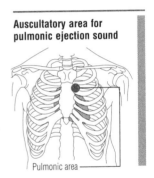

Auscultatory area for pulmonic ejection sound

Pulmonic area

PCG and ECG showing pulmonic ejection sound

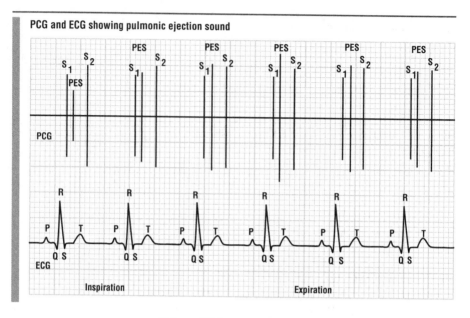

Inspiration   Expiration

Using a PES to differentiate clinical conditions

Because a PES may occur in idiopathic dilation of the pulmonary artery, but usually isn't present in supravalvular or muscular subvalvular obstructions, its presence can be used to differentiate between these conditions. Occasionally, a PES may also be heard with atrial and ventricular septal defects.

AORTIC EJECTION SOUNDS

An aortic ejection sound (AES) is heard best near the heart's apex over the mitral area, near the heart's base over the aortic area, or over Erb's point. (◆1-39) An AES may be heard in patients with aortic stenosis or aortic insufficiency, but the sound is less clicklike when associated with aortic insufficiency.

Unlike a PES, caused by pulmonic valve stenosis, an AES, caused by aortic valve stenosis, doesn't vary in intensity with respiration. An AES may occur in patients with aortic root dilation, which is commonly associated with such conditions as systemic hypertension, an ascending aortic aneurysm, or coarctation of the aorta.

Sound characteristics

An AES is heard best near the heart's apex over the mitral area, near the heart's base over the aortic area, or over Erb's point. Its intensity is soft but may be equal to or greater

Auscultatory area for aortic ejection sound

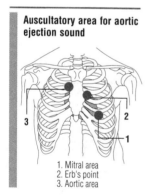

1. Mitral area
2. Erb's point
3. Aortic area

PCG and ECG showing aortic ejection sound

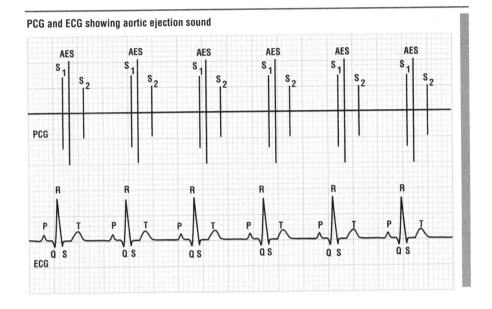

than that of S_1. An AES has a short duration. It has a high pitch that's heard best using the diaphragm of the stethoscope and a sharp, or clicklike, quality. Its timing is closely related to S_1. (♦1-40) An AES occurs early in ventricular systole, just after the opening of a stenotic aortic valve, and is heard just after the QRS complex in the ECG waveform.

Differentiating an AES from other heart sounds
Certain characteristics help to distinguish an AES from other heart sounds. One characteristic is that an AES radiates more than a PES. Another characteristic is that a split S_1 heard over the mitral area is likely to be an M_1 and an AES.

You can differentiate an AES from an S_4 by remembering that S_4 is heard best over the mitral area using the bell of the stethoscope and is commonly accompanied by a palpable, presystolic apical bulge. Also, an S_4 is intensified by maneuvers that increase left atrial pressure, such as brief exercise, squatting, or coughing. An AES is unaffected by these maneuvers.

CLICKS
A midsystolic click (MSC) occurs when the prolapsed mitral valve's leaflets and chordae tendineae tense. The anterior leaflet, the posterior leaflet, or both can prolapse. Occasionally, multiple clicks occur that are heard in midsystole to late systole; they're heard best over the tricuspid area and

Area producing midsystolic click

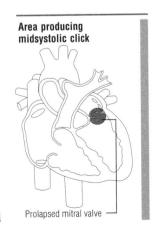

Prolapsed mitral valve

PCG and ECG showing midsystolic click

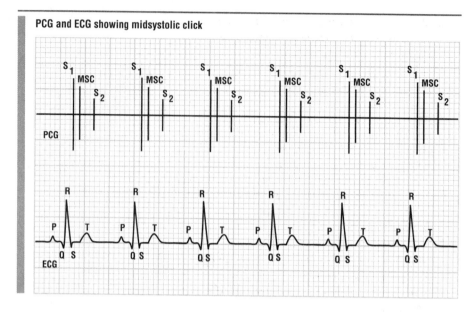

toward the mitral area. Like ESs, these midsystolic to late-systolic clicks are crisp, high-frequency sounds. (◆1-41)

Sound characteristics

An MSC is usually heard best over the tricuspid area and near the heart's apex over the mitral area. Its intensity is equal to or greater than that of S_1. An MSC has a short duration. It has a high pitch that's heard best using the diaphragm of the stethoscope, and its quality is clicklike. The click's timing varies; it can occur in early systole, midsystole, or late systole. An MSC can be heard during the QT interval in the ECG waveform. (◆1-42)

Auscultatory area for midsystolic click

1. Tricuspid area
2. Mitral area

Differentiating an MSC from other heart sounds

The timing of an MSC is affected by various maneuvers, such as having the patient stand or perform Valsalva's maneuver. Such maneuvers result in reduced left ventricular filling and cause the MSC to be heard closer to S_1. The MSC may even merge with S_1 and disappear completely. Increasing left ventricular volume by raising the legs from a recumbent position or squatting delays the click. This maneuver may also cause the prolapse not to occur and the MSC to be diminished or inaudible.

Sometimes an MSC is accompanied by a late systolic crescendo murmur, as in mitral valve prolapse, or the characteristic holosystolic murmur, as in mitral regurgitation.

An MSC may also be caused by some extracardiac conditions, such as pleuropericardial adhesions, atrial septal aneurysms, and cardiac tumors.

◀◀◀ POSTTEST

1. What heart sound is produced when the mitral valve leaflets become stenotic or abnormally narrowed?
2. Describe the characteristics of an OS.
3. How can you tell the difference between an OS and S_2?
4. How can you tell the difference between an OS and S_3?
5. What is a systolic ejection sound?
6. What causes a PES?
7. Describe the characteristics of a PES.
8. What is the primary difference between a PES and an AES?
9. Describe the characteristics of an AES.
10. How can you tell the difference between an AES and other heart sounds?
11. What causes an MSC?
12. Describe the characteristics of an MSC.
13. How can you tell the difference between an MSC and other heart sounds?

Murmur fundamentals

7

PRETEST ▶▶▶

1. How does turbulent blood flow produce murmurs?
2. What are the seven characteristics used to describe murmurs?
3. What is the six-point graded scale?
4. What is a murmur's configuration?

Turbulent blood flow

Whereas heart sounds are produced by brief vibrations that correspond to the beginning and end of systole, murmurs are produced by a prolonged series of vibrations that occurs during systole, diastole, or both. These vibrations result from turbulent blood flow.

Several clinically significant conditions—such as blood flowing at a high velocity through a partially obstructed opening, blood flowing from a higher pressure chamber to a lower pressure one, or any combination of these—can cause turbulent blood flow.

❀ **AGE ISSUE** *Innocent murmurs, such as Still's murmur, are commonly heard in young children; they typically disappear in puberty. Pathognomonic heart murmurs in infants and young children usually result from congenital heart disease, such as atrial and ventricular septal defects. Other murmurs heard in pediatric patients can be acquired, such as with rheumatic heart disease.*

Characteristics

Murmurs, like other heart sounds, are described by their audible characteristics heard during auscultation. The terms used to describe a specific characteristic are determined primarily by the volume and speed of the jet of blood as it moves through the heart.

The seven characteristics used to describe murmurs are *location, intensity, duration, pitch, quality, timing,* and *configuration.*

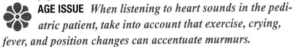 **AUSCULTATION TIP** *Initially, learn to identify the location and timing of murmurs. As your auscultation techniques improve, try to identify the intensity, duration, pitch, quality, and configuration.*

AGE ISSUE *When listening to heart sounds in the pediatric patient, take into account that exercise, crying, fever, and position changes can accentuate murmurs.*

LOCATION

A murmur's location is the anatomic area on the chest wall where the murmur is heard best and is usually also the murmur's point of maximum intensity. This area usually correlates with the underlying location of the valve that's responsible for producing the murmur. For example, an aortic stenosis murmur is usually heard best near the heart's base over the aortic area, whereas a mitral regurgitation murmur is usually heard best near the heart's apex over the mitral area.

The murmur's sounds may also be transmitted to the chamber or vessel where the turbulent blood flow occurs. This phenomenon, known as *radiation,* occurs because the direction of blood flow determines sound transmission. Murmurs radiate in either a forward or a backward direction.

INTENSITY

The second characteristic, intensity, refers to the murmur's loudness. Intensity is influenced by many factors, including body weight. For example, because transmission of heart sounds to the chest wall is affected by chest wall thickness and by certain diseases, heart sounds and murmurs are usually louder in thin individuals and fainter in obese individuals. They're also fainter in patients with emphysema.

Hyperdynamic states, decreased blood viscosity, increased pressure gradients across valves, larger jets of blood, and faster heart rates may also increase a murmur's intensity. Murmurs are less intense in hypodynamic states and in patients with an elevated hematocrit.

Documenting intensity

Document a murmur's intensity using a uniform method. Most health care professionals use a six-point graded scale, with 1 being the faintest intensity and 6 being the loudest. A grade 1 murmur is faint, may be heard intermittently, and is barely heard through the stethoscope. A grade 2 murmur is also faint but is usually heard as soon as the stethoscope is placed on the chest wall. A grade 3 murmur is easily heard and is described as moderately loud. A grade 4 murmur is loud and is usually associated with a palpable vibration known as a *thrill*. It also may radiate in the direction of blood flow. A grade 5 murmur is loud enough to be heard with only an edge of the stethoscope touching the chest wall; it's almost always accompanied by a thrill and radiation. A grade 6 murmur is so loud that it can be heard with the stethoscope close to, but not touching, the chest wall; it's always accompanied by a thrill, and it radiates to distant structures.

When you document a murmur's grade, write it as a fraction — for example, ⅜. This means you consider it to be a grade 3 and have used the six-point scale for assessment. That way, all health care professionals will understand which scale was used, even if they don't use the same one.

DURATION

Duration, the length of time the murmur is heard during systole or diastole, may be described as long or short.

PITCH

A murmur's pitch, or frequency, varies from high to low. It's usually higher in conditions accompanied by increased blood flow velocity or increased pressure gradients and lower in conditions associated with lower blood flow velocity.

QUALITY

A murmur's quality is determined by the combination of frequencies that produces the sound. It's typically described as harsh, rough, musical, scratchy, squeaky, rumbling, or blowing.

TIMING

A murmur's timing refers to when the murmur occurs in the cardiac cycle. This means that the onset, duration, and end of the murmur are described in relation to systole and diastole. The beginning of systole, or the first heart sound (S_1), can be identified easily by palpating the carotid pulse or by looking for the QRS complex on the electrocardiogram (ECG) monitor's oscilloscope.

All systolic murmurs occur between S_1 and the second heart sound (S_2) during ventricular systole; this interval is between the QRS complex and the T wave in the ECG waveform. All diastolic murmurs occur between S_2 and S_1 during ventricular diastole; this interval is between the T wave and the QRS complex in the ECG waveform.

Murmurs are further classified according to their timing within the phases of the cardiac cycle. For example, a murmur can be described as holosystolic, meaning it's present throughout systole; early systolic, midsystolic, or late systolic; or early diastolic, middiastolic, or late diastolic.

CONFIGURATION

The last characteristic, configuration, refers to the shape of a murmur's sound as recorded on a phonocardiogram. The configuration is usually defined by changes in the murmur's intensity during systole or diastole and is determined by blood flow pressure gradients. For example, a *crescendo murmur* is one that gradually increases in intensity as the pressure gradient increases. A *decrescendo murmur* is one that gradually decreases in intensity as the pressure gradient decreases. A *crescendo-decrescendo murmur* first increases in intensity as the pressure gradient increases, then decreases in intensity as the pressure gradient decreases; it's also known as a *diamond-shaped murmur.* Finally, a *plateau-shaped murmur* is one that's equal in intensity throughout the murmur.

PUTTING THEM ALL TOGETHER

Murmurs must be described carefully and accurately using these seven characteristics. Making such information immediately available to all who care for a specific patient allows any changes in a murmur's characteristics to be recognized easily. In that way, the possible source of those changes can be assessed more easily.

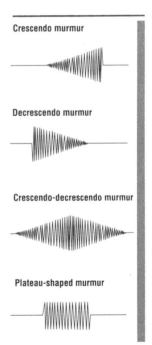

Crescendo murmur

Decrescendo murmur

Crescendo-decrescendo murmur

Plateau-shaped murmur

◀◀◀ **POSTTEST**

1. What causes turbulent blood flow?
2. What determines a murmur's sound characteristics?
3. List the seven characteristics used to describe murmurs.
4. What is a murmur's location?
5. What does the term *radiation* mean?
6. What is a murmur's intensity?
7. Describe the six-point graded scale.
8. What is a murmur's duration?
9. What is a murmur's pitch?
10. What is a murmur's quality?
11. What is a murmur's timing?
12. What is a murmur's configuration?
13. What are the four configurations used to describe a murmur?

Systolic murmurs

8

PRETEST ▶▶▶

1. How is the sound of a systolic murmur produced?
2. What are ventricular outflow obstruction murmurs?
3. List the murmurs associated with right ventricular outflow obstruction.
4. List the murmurs associated with left ventricular outflow obstruction.
5. How is the sound of a systolic regurgitation murmur produced?
6. List the systolic regurgitation murmurs.

Anatomy and physiology

Normally, as ventricular pressures rise at the beginning of systole, the mitral and tricuspid valves close. Then, for a brief time, while the aortic and pulmonic valves are still closed during isovolumic contraction, ventricular pressures rise sharply. When the pressure in both ventricles is high enough, the aortic and pulmonic valves open, and blood is ejected from the ventricles into the aorta and the pulmonary artery. Normally, functioning valves facilitate this unidirectional blood flow.

However, aortic or pulmonic outlet abnormalities may generate forward systolic ejection murmurs. When the mitral or tricuspid valve is involved, backward, or *regurgitant*, murmurs may be heard during systole. All systolic murmurs occur during ventricular systole between the first and second heart sounds (S_1 and S_2).

✤ **AGE ISSUE** *Between ages 30 and 80, the left ventricular wall grows 25% thicker from its increased efforts to pump blood. Heart valves also become thicker from fibrotic* *and sclerotic changes. This thickening can prevent the valves from closing completely, causing systolic murmurs.*

Systolic ejection murmurs

Area producing systolic ejection murmur

1. Aortic valve
2. Pulmonic valve
3. Right ventricle
4. Left ventricle

During ventricular systole, the rapid ejection of blood from the ventricles causes turbulent blood flow, which produces an innocent systolic ejection murmur (SEM). (♦1-43) This murmur, which is considered to be normal, is usually inaudible. However, in thin chested individuals, children, and patients with hyperdynamic states, such as anemia, thyrotoxicosis, fever, pregnancy, or arteriovenous shunts, an SEM can usually be auscultated.

An SEM, classified as benign or functional, is called an *innocent* or *flow murmur* because there's no associated physiologic or structural abnormality. However, unusual turbulence related to dilation of the aorta or pulmonary artery may also produce an innocent SEM. Researchers believe SEMs can be caused by sclerotic changes in the aortic valve cusps.

AGE ISSUE *About 50% of all persons older than age 50 have this type of murmur, which is heard more commonly in women and in patients with hypertension.*

Sound characteristics

An SEM is usually heard best along the left sternal border and sometimes over the aortic and mitral areas. Its intensity is usually soft (less than a grade ⅜), and the duration of the murmur is short.

AUSCULTATION TIP *To determine if a murmur intensifies, listen at the murmur's border.*

An SEM has a medium pitch that's heard best using the diaphragm of the stethoscope. The quality of an SEM varies. Its timing is early systolic; it ends well before a normal S_2 split. It's heard after the QRS complex on the electrocardiogram (ECG) waveform. An SEM has a crescendo-decrescendo configuration. (♦1-44)

Auscultatory area for systolic ejection murmur

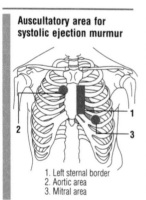

1. Left sternal border
2. Aortic area
3. Mitral area

Enhancement techniques

An SEM's intensity can be increased by maneuvers that increase blood volume or ejection velocity, such as having the patient raise his legs from a recumbent position, exercise briefly, or cough a few times.

PCG and ECG showing systolic ejection murmur

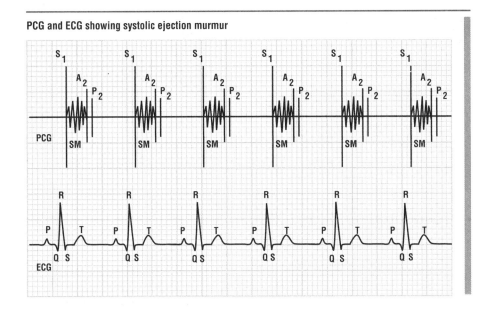

Ventricular outflow obstruction murmurs

Right or left ventricular outflow obstructions may be supravalvular, valvular, or subvalvular. Regardless of location, the outflow obstruction causes turbulent blood flow, which produces a midsystolic ejection murmur. The murmur begins early in systole — after S_1 and the opening of the diseased pulmonic or aortic valve. It ends before the S_2 closure component of the diseased pulmonic or aortic valve. This murmur typically has a crescendo-decrescendo configuration that peaks in intensity in early systole, midsystole, or late systole, depending on the severity of the obstruction.

RIGHT VENTRICULAR OUTFLOW OBSTRUCTION MURMURS
Supravalvular pulmonic stenosis murmurs
Supravalvular pulmonic stenosis, or *pulmonary artery branch stenosis,* is a type of right ventricular outflow obstruction that occurs above the pulmonic valve. This obstruction produces a murmur with a crescendo-decrescendo configuration that's occasionally continuous. (**♦1-45**) The murmur is rarely accompanied by a pulmonic ejection sound (PES) and is associated with a P_2 of normal intensity.

Area producing supravalvular pulmonic stenosis murmur

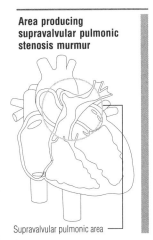

Supravalvular pulmonic area —

PCG and ECG showing supravalvular pulmonic stenosis murmur

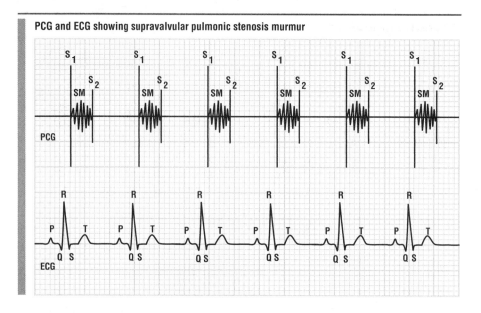

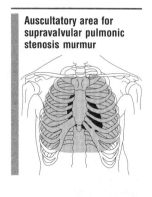

Auscultatory area for supravalvular pulmonic stenosis murmur

Sound characteristics

The supravalvular pulmonic stenosis murmur is usually heard over much of the thorax. Its intensity and duration vary. It has a medium pitch that's heard best using the diaphragm of the stethoscope, and it has a harsh quality. Its timing is systolic: It begins after S_1 and ends before a normal S_2 split. On the ECG waveform, the murmur begins just after the QRS complex begins and ends just before the T wave ends. It has a crescendo-decrescendo configuration that's occasionally continuous. (◆1-46)

Pulmonic valvular stenosis murmurs

The pulmonic valvular stenosis murmur results from congenital pulmonic valvular stenosis and is commonly associated with other congenital defects. In mild pulmonic valvular stenosis, S_1 is normal. The murmur begins after S_1 with a right-sided PES as the pulmonic valve abruptly stops opening. Remember, the PES is the only right-sided heart sound that increases in intensity during expiration and becomes less audible during inspiration. The murmur intensifies after the PES and peaks in midsystole; then it begins to fade. It ends before S_2. The intensity of P_2 is normal. (◆1-47)

In severe pulmonic valvular stenosis, the pressure gradient across the pulmonic valve increases. An increased pressure gradient causes the PES to be heard earlier; it may

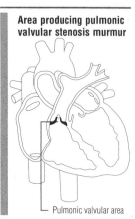

Area producing pulmonic valvular stenosis murmur

Pulmonic valvular area

PCG and ECG showing pulmonic valvular stenosis murmur

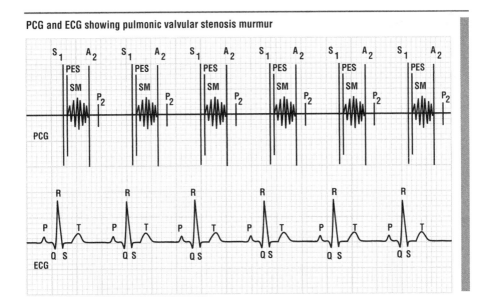

even fuse with S_1. Right ventricular ejection time is also prolonged. Consequently, the murmur has a longer crescendo and the intensity peaks later in systole. It continues throughout A_2 but ends before P_2. Prolonged right ventricular ejection time also causes a delayed P_2, creating a wide S_2 split.

Usually, as the stenosis becomes more severe, the murmur's duration lengthens and its configuration becomes more asymmetrical. Consequently, P_2 is delayed even more and its intensity is decreased.

Sound characteristics

The pulmonic valvular stenosis murmur is heard best near the heart's base over the pulmonic area. It commonly radiates toward the left side of the neck or the left shoulder. The murmur's intensity is usually soft, but it will become louder with a palpable thrill toward the left side of the neck and the left shoulder as the stenosis becomes more severe. Its duration is short, but this too will increase as the stenosis worsens. It has a medium pitch that's heard best using the diaphragm of the stethoscope and a harsh quality. Its timing is midsystolic: It ends before a normal S_2 split. It's accompanied by a PES that diminishes or disappears with inspiration. On the ECG waveform, it begins after the QRS complex and ends before the T wave ends. The murmur has a crescendo-decrescendo configuration. A mild pulmonic

Auscultatory area for pulmonic valvular stenosis murmur

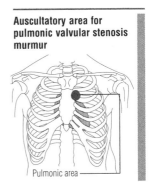

Pulmonic area

PCG and ECG showing subvalvular pulmonic stenosis murmur

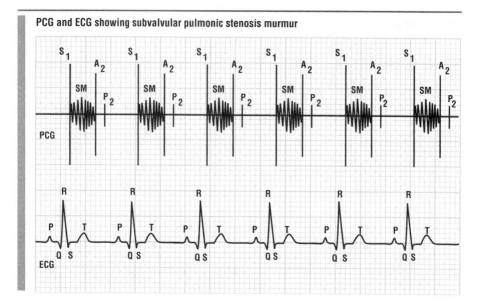

Area producing subvalvular pulmonic stenosis murmur

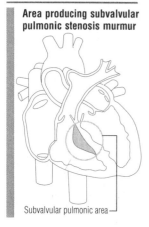

Subvalvular pulmonic area

Auscultatory area for subvalvular pulmonic stenosis murmur

1. Pulmonic area
2. Erb's point

valvular stenosis murmur is shaped like a diamond; a severe pulmonic valvular stenosis murmur, like a kite. **(◆1-48)**

Subvalvular pulmonic stenosis murmurs

When a right ventricular outflow obstruction is beneath the pulmonic valve, or subvalvular, the midsystolic ejection murmur sounds the same as a pulmonic valvular stenosis murmur. However, it isn't initiated by a PES. **(◆1-49)** If the subvalvular obstruction is associated with a ventricular septal defect, the murmur will be more complex in character.

Sound characteristics

The subvalvular pulmonic stenosis murmur is usually heard best over the pulmonic area and over Erb's point; it commonly radiates toward the left side of the neck, the left shoulder, or both. Its intensity is usually soft but becomes louder as the stenosis worsens. The duration, which is short, also increases as the stenosis worsens. The murmur has a medium pitch that's heard best using the diaphragm of the stethoscope; its quality is harsh. Its timing is midsystolic: It starts after S_1 and ends before a normal S_2 split. On an ECG, the murmur begins after the QRS complex and ends before the T wave ends. **(◆1-50)** This murmur isn't initiated by a PES. It has a crescendo-decrescendo configuration.

PCG and ECG showing supravalvular aortic stenosis murmur

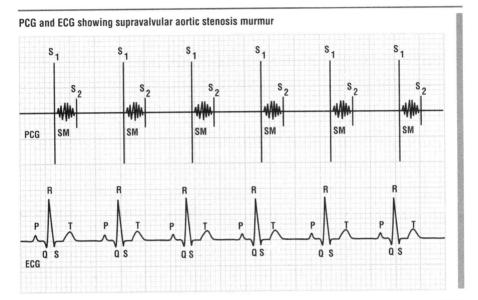

LEFT VENTRICULAR OUTFLOW OBSTRUCTION MURMURS

Supravalvular aortic stenosis murmurs

When a left ventricular outflow obstruction is above the aortic valve, or supravalvular, it produces a murmur with a crescendo-decrescendo configuration. (◆1-51) Supravalvular aortic stenosis is usually congenital. Its characteristics are similar to those of aortic valvular stenosis, except that it has no aortic ejection sound (AES), is louder, and is heard best over the suprasternal area, aortic area, or right first intercostal space.

Sound characteristics

The supravalvular aortic stenosis murmur is usually heard best near the heart's base over the right first intercostal space, over the aortic area, and over the suprasternal notch. It may radiate toward the right side of the neck, the right shoulder, or both. Its intensity, typically a grade ⅜ to ⅘, decreases in patients with left-sided heart failure. Its duration increases as the stenosis worsens. It has a medium pitch that can be heard equally well with the diaphragm or bell of the stethoscope. The murmur has a rough quality. Its timing is midsystolic: It ends before a normal S_2 split. It isn't associated with an AES. On the ECG waveform, it begins after the QRS complex and ends before the T wave ends. The

Area producing supraval-vular aortic stenosis murmur

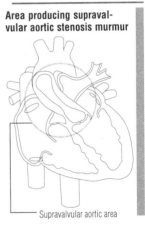

Supravalvular aortic area

Auscultatory area for supraval-vular aortic stenosis murmur

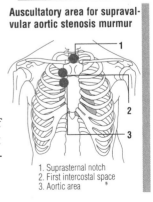

1. Suprasternal notch
2. First intercostal space
3. Aortic area

murmur has a crescendo-decrescendo configuration. **(◆1-52)**

Area producing aortic valvular stenosis murmur

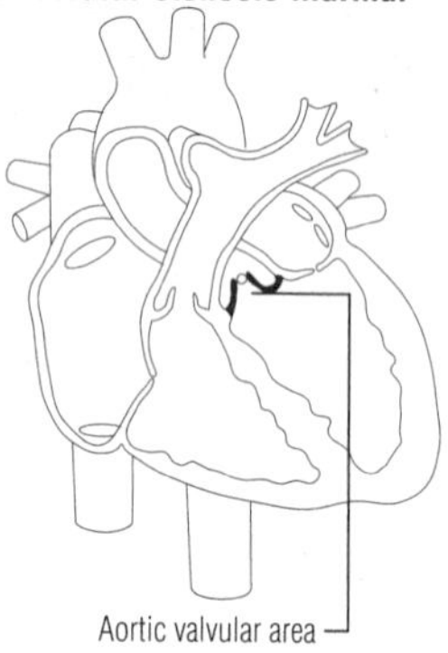

Aortic valvular stenosis murmurs

Aortic valvular stenosis may be congenital or may be acquired from degenerative or rheumatic heart disease. If it's acquired from rheumatic heart disease, the mitral valve is commonly affected as well. Aortic stenosis produces an SEM that begins after S_1 and ends before S_2. **(◆1-53)** After S_1, left ventricular pressure rises. The stenotic aortic valve halts its opening motion and produces a loud AES that's heard best near the heart's apex over the mitral area. The AES is followed by the murmur, which gradually intensifies until midsystole to late systole and then fades, ending before a faint A_2.

AGE ISSUE ***As a person ages, such changes as the development of fibrous tissues, which thicken the bases of the aortic leaflets, and calcification can lead to an aortic systolic murmur. Aortic stenosis develops if the valve leaflets become increasingly stiff and stationary. This is a gradual process that occurs across the lifespan.***

When an aortic valvular stenosis murmur is discovered late in adulthood, it can be heard best over the mitral area. As the valve becomes more calcified with age—and consequently less mobile—both the AES and A_2 become soft or even inaudible. The murmur can vary in intensity from a soft grade 2/6 to a rough grade 4/6 sound, but most aortic valvular stenosis murmurs are a grade 3/6 to 4/6.

Auscultatory area for aortic valvular stenosis murmur

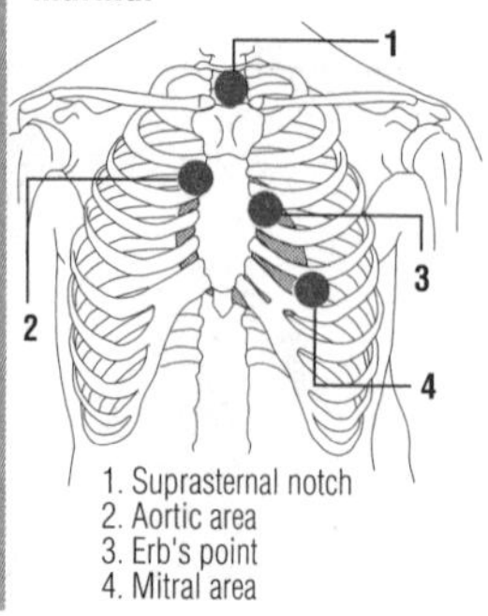

Sound characteristics

The aortic valvular stenosis murmur is usually heard best near the heart's base over the aortic area, over Erb's point, near the heart's apex over the mitral area, or over the suprasternal notch. The murmur may radiate toward the right side of the neck, the right shoulder, or both. A thrill may be palpable over the aortic area and neck. The murmur's intensity, typically a grade 3/6 to 4/6, decreases in patients with left-sided heart failure. Its duration increases as stenosis worsens.

The aortic valvular stenosis murmur has a medium pitch that's heard equally well using the diaphragm or the bell of the stethoscope. The murmur's quality is rough and may become harsher and louder as stenosis worsens. Its timing is midsystolic. An AES, when present, is heard shortly after S_1; the AES is followed by the murmur, which

PCG and ECG showing aortic valvular stenosis murmur

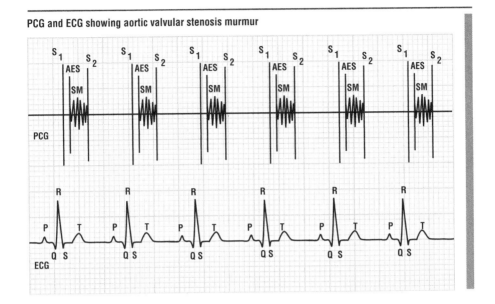

ends before a normal S_2 split. If an S_4 is heard before age 40, stenosis is usually severe. On the ECG waveform, the murmur begins after the QRS complex and ends before the T wave ends. It has a crescendo-decrescendo configuration. (◆1-54)

Variations
Certain characteristics of the aortic valvular stenosis murmur vary according to the degree of stenosis. For example, as stenosis worsens, left ventricular ejection time is prolonged, and the murmur peaks later in systole. A delayed A_2 and a shortened A_2-P_2 interval are also heard.

Also, the more severe the stenosis, the closer A_2 moves to P_2. With extremely severe stenosis, A_2 becomes soft and actually follows P_2, producing a paradoxical S_2 split.

Subvalvular aortic stenosis murmurs

When a left ventricular outflow obstruction is below the aortic valve, or subvalvular, it produces a murmur with a crescendo-decrescendo configuration. (◆1-55) A subvalvular aortic outflow obstruction may be caused by a congenital fibrous ring, or it may be acquired from hypertrophic obstructive cardiomyopathy (formerly known as *idiopathic hypertrophic subaortic stenosis*). This is a dynamic obstruction produced by asymmetrical hypertrophy of the septum

Area producing subvalvular aortic stenosis murmur

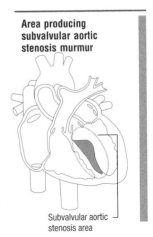

Subvalvular aortic stenosis area

PCG and ECG showing subvalvular aortic stenosis murmur

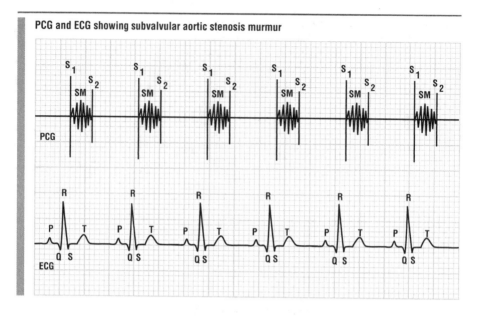

and an abnormal anterior motion of the mitral valve leaflets during systole.

Sound characteristics

The subvalvular aortic stenosis murmur is usually heard best near the heart's apex over the mitral and tricuspid areas. It doesn't usually radiate toward the base, right side of the neck, or right shoulder. Its intensity, typically a grade ⅜ to ⅚, increases as stenosis worsens; its duration varies. The murmur has a medium pitch that's heard equally well using the bell or the diaphragm of the stethoscope. Its quality can be harsh or rough. This murmur's timing is midsystolic: It peaks in midsystole and ends before a normal S_2 split, a delayed A_2, or a paradoxical S_2 split. On the ECG waveform, it begins after the QRS complex and ends before the T wave ends. The murmur has a crescendo-decrescendo configuration. (◆1-56)

Differentiating subvalvular and aortic valvular stenosis murmurs

A subvalvular hypertrophic cardiomyopathic murmur becomes louder during Valsalva's maneuver, whereas an aortic valvular stenosis murmur doesn't. Performing Valsalva's maneuver or having the patient stand up suddenly decreases venous return and left ventricular filling; this makes the left ventricle smaller, the obstruction more severe, and the sub-

Auscultatory area for subvalvular aortic stenosis murmur

1. Tricuspid area
2. Mitral area

valvular hypertrophic obstructive cardiomyopathic murmur louder. Having the patient squat increases peripheral vascular resistance and left ventricular filling; this maneuver decreases the pressure gradient across the aortic valve and decreases or obliterates the subvalvular hypertrophic obstructive cardiomyopathic murmur.

Systolic regurgitation murmurs

An abnormality of either the tricuspid or mitral valve may result in backward turbulent blood flow during systole. This means that blood moves in a direction opposite that of the normal unidirectional flow pattern. Blood regurgitates through a defective, incompetent tricuspid or mitral valve into the left or right atrium, resulting in what is commonly called tricuspid or mitral valve insufficiency. An incompetent valve may be caused by a primary valvular disorder or may develop secondary to dysfunction of the valve's supporting structures.

The regurgitant murmurs heard in patients with these disorders may be early or late systolic, or holosystolic. The early systolic murmurs have a crescendo-decrescendo configuration; the late systolic murmurs, either a crescendo or crescendo-decrescendo configuration; and the holosystolic murmurs, a plateau shape.

TRICUSPID REGURGITATION MURMURS
The tricuspid regurgitation murmur, also called *tricuspid insufficiency*, most commonly results from right ventricular dilation, which usually results from mitral valve disease or left-sided heart failure but may also be caused by pulmonary disease. It occasionally results from a congenital valve malformation and may also be acquired as a result of infective endocarditis, a right ventricular infarction, or trauma to the valve or its supporting structures.

This murmur can also be heard occasionally in a patient with a transvenous pacemaker because the pacemaker lead interferes with tricuspid valve closure, creating a tricuspid regurgitation type murmur. The T_1 (tricuspid valve closure) intensity is either increased, normal, or decreased. (**♦1-57**)

Sound characteristics
The tricuspid regurgitation murmur is usually heard best over the tricuspid area. In some patients, it's heard only dur-

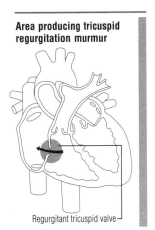

Area producing tricuspid regurgitation murmur

Regurgitant tricuspid valve

PCG and ECG showing tricuspid regurgitation murmur

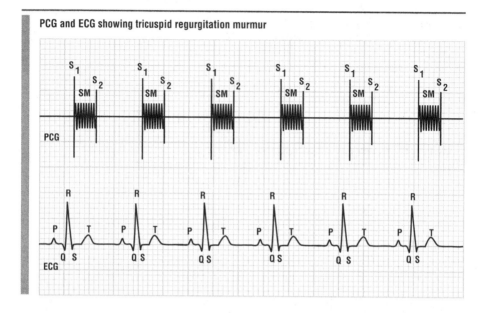

Auscultatory area for tricuspid regurgitation murmur

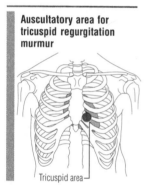

Tricuspid area

ing inspiration. The murmur may radiate to the right of the sternum. Its usual soft intensity may increase during deep inspiration. The murmur's duration is long. It has a medium pitch that's heard best using the diaphragm of the stethoscope and a scratchy or blowing quality. The murmur's timing is systolic: It lasts from S_1 to P_2. On the ECG waveform, the murmur begins just after the QRS complex and ends after the T wave. It's holosystolic and plateau shaped. (◆1-58)

Enhancement techniques

Regardless of the tricuspid regurgitation murmur's cause, the murmur increases in intensity during inspiration because venous return and right ventricular filling are increased, creating higher pressure gradients during systole. To enhance the murmur, have the patient breathe through his mouth slowly, quietly, and more deeply while sitting or standing. Caution the patient not to hold his breath because this negates the maneuver's effect. Another technique is to apply pressure inward and upward below the right costal margin. The murmur is louder during inspiration. In some patients, this is the only time the murmur is audible. The inspiratory increase in loudness of right-sided auscultatory events is called *Carvallo's sign*.

If the tricuspid regurgitation murmur is secondary to right-sided heart failure, right ventricular filling may be

PCG and ECG showing mitral regurgitation murmur

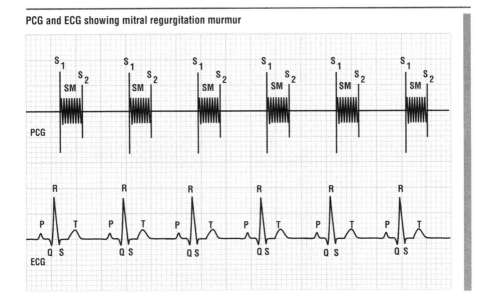

limited with inspiration; consequently, the murmur may not be intensified during deep breathing.

This murmur is rarely accompanied by a systolic thrill, and neither its timing in systole nor its intensity correlates with the severity of the regurgitation.

MITRAL REGURGITATION MURMURS

The mitral regurgitation murmur, also called *mitral insufficiency,* may be caused by a congenital or acquired abnormality of the mitral valve leaflets, the valve's supporting structures, or the left ventricle. An incompetent mitral valve causes backward blood flow through the incompetent mitral valve during systole. The increased pressure at the aortic valve facilitates regurgitation of blood through the incompetent mitral valve into the low-pressure left atrium, producing the mitral regurgitation murmur. (◆1-59)

This murmur may appear in early systole, midsystole, or late systole, or it may be holosystolic. Its quality, duration, and radiation depend on the extent, duration, and location of the disease. Usually, late systolic murmurs are associated with mild to moderate regurgitation.

 AGE ISSUE *As a person ages, fibrotic and sclerotic changes thicken the mitral valve leaflets, leading to rigidity and incomplete closure.*

Area producing mitral regurgitation murmur

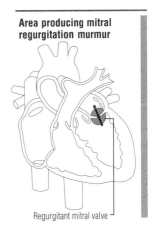

Regurgitant mitral valve

Holosystolic mitral regurgitation murmurs

The holosystolic mitral regurgitation murmur can be caused by the effects of rheumatic heart disease on the mitral valve, by mitral valve prolapse, by left ventricular dilation, or by papillary muscle dysfunction. The murmur can usually be heard regardless of the patient's position, but grade $\frac{1}{6}$ to $\frac{2}{6}$ murmurs may be heard better with the patient in the partial left lateral recumbent position or after exercise.

A loud S_3 usually accompanies moderate to severe mitral regurgitation; this sound is related not to heart failure but to the increased left ventricular volume in early diastole.

 AGE ISSUE *A systolic or continuous murmur called mammary souffle may be heard in women in the late stages of pregnancy or early in the postpartum period. It's caused by increased breast circulation. Typically, firm pressure to the diaphragm of the stethoscope terminates the murmur. It's heard best with the patient in the supine position on the left or right second or third intercostal spaces.*

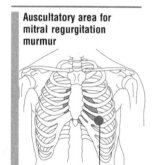

Auscultatory area for mitral regurgitation murmur

Mitral area

Sound characteristics

The holosystolic mitral regurgitation murmur is usually heard best near the heart's apex over the mitral area. The sound may radiate to the axillae or posteriorly over the lung bases. Its intensity is variable and is usually unaffected by respiration; however, it may be somewhat diminished during inspiration. The murmur has a long duration. It has a medium to high pitch that's heard best using the diaphragm of the stethoscope. The murmur may be accompanied by a systolic apical thrill. Its quality is blowing, and its timing is systolic — from S_1 to S_2. On the ECG waveform, it appears just after the QRS complex to the end of the T wave. The murmur is holosystolic and plateau shaped. (**◆1-60**)

Acute mitral regurgitation murmurs

Acute mitral regurgitation is less common than chronic mitral regurgitation. Its murmur results from rupture of the chordae tendineae, papillary muscle, or both. Occasionally, this murmur occurs with a myocardial infarction, but it may also result from severe damage to the mitral valve from trauma or infection.

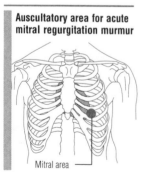

Auscultatory area for acute mitral regurgitation murmur

Mitral area

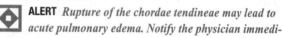 **ALERT** *Rupture of the chordae tendineae may lead to acute pulmonary edema. Notify the physician immediately, and monitor the patient for signs and symptoms of acute pulmonary edema (acute shortage of breath; rapid respiration; audible wheezes and crackles; productive cough that produces*

PCG and ECG showing acute mitral regurgitation murmur

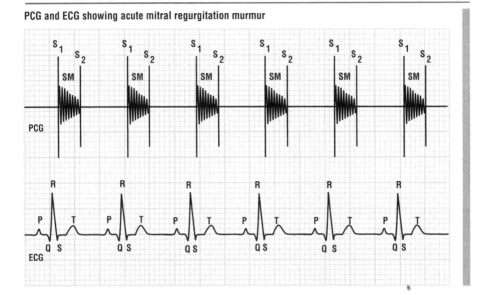

frothy, blood-tinged sputum; tachycardia; hypotension; and cyanotic, cold, clammy skin).

The acute mitral regurgitation murmur begins with mitral valve closure and is decrescendo. (**◆1-61**) Its intensity decreases as ventricular and atrial pressures equilibrate during late systole. An S_4 is usually heard in acute mitral regurgitation.

Sound characteristics

The acute mitral regurgitation murmur is usually heard best near the heart's apex over the mitral area. Its intensity is usually loud (grade ⅘ to ⅚ if the murmur results from rupture of the chordae tendineae). It's accompanied by a systolic thrill. The murmur's duration is medium long. It has a high pitch that's heard best using the diaphragm of the stethoscope and a quality that can be musical. Its timing is systolic: It begins with M_1 and ends with or before A_2. On the ECG waveform, the murmur begins just after the QRS complex and ends just after the T wave. In many patients, it's holosystolic and wedge shaped (the wedge shape has a steeper decrescendo configuration). (**◆1-62**)

Mitral valve prolapse murmurs

The murmur associated with a prolapsed mitral valve usually appears in late systole and is either isolated or accompanied by a nonejection midsystolic click or clicks caused by

Area producing mitral valve prolapse murmur

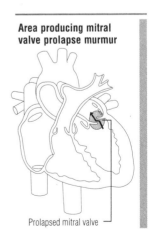

Prolapsed mitral valve

PCG and ECG showing mitral valve prolapse murmur

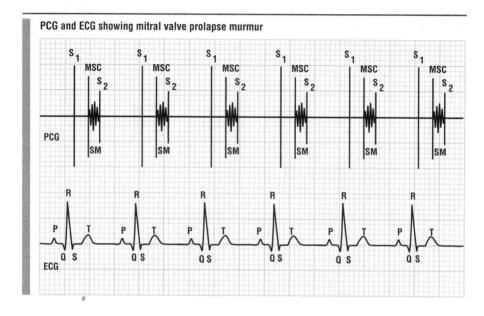

the mitral valve prolapsing and ballooning up into the left atrium. This is sometimes referred to as the click murmur syndrome. (◆1-63)

🏵 **AGE ISSUE** *Mitral valve prolapse is one of the most common valvular abnormalities found in adults.*

Auscultatory area for mitral valve prolapse murmur

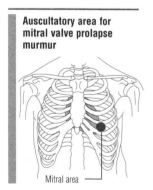

Mitral area

Sound characteristics

The mitral valve prolapse murmur is usually heard best near the heart's apex over the mitral area. Its intensity is usually soft (grade ⅖ to ⅜), and its duration is short. It has a high pitch that's heard best using the diaphragm of the stethoscope. The murmur has a musical quality; when loud, it's sometimes described as a whoop or honk. Although its timing is usually late systolic, it can sometimes be holosystolic. On the ECG waveform, the murmur coincides with the T wave and ends just after the T wave. It has a crescendo or crescendo-decrescendo configuration. (◆1-64)

Enhancement techniques

Having the patient stand decreases left ventricular volume, which causes the mitral valve prolapse murmur to be heard earlier in systole, to be louder, and to last longer. Having the patient squat increases left ventricular volume, which causes the murmur to be heard later in systole.

The direction in which the mitral valve prolapse murmur is transmitted across the chest wall depends on the dis-

PCG and ECG showing a ventricular septal defect murmur

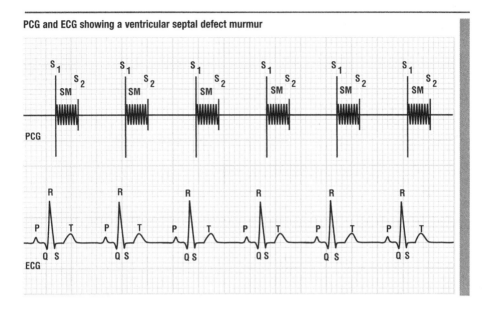

ease. For example, if the mitral valve's posterior leaflet is involved, blood flow might be directed more anteriorly and medially. Consequently, the murmur would be transmitted toward the heart's base and heard best over the aortic area and along the left sternal edge. If the anterior leaflet is incompetent, blood flow is directed posteriorly against the posterior left atrial wall, and the murmur is transmitted toward the left axilla and back.

Ventricular septal defect murmur

A ventricular septal defect (VSD) is an opening — usually in the membranous portion of the ventricular septum — that allows direct communication between the ventricles. Typically, blood shunts from left to right because of higher pressures in the left ventricle.

 AGE ISSUE *Murmurs in pediatric patients are typically associated with ventricular septal defects, patent ductus arteriosus, atrial septal defects, and mitral valve prolapse.*

Sound characteristics
The VSD murmur is heard over the lower sternal border and, if loud, can be heard over the entire precordium. A thrill may be palpable along the lower left sternal border.

Area producing ventricular septal defect murmur

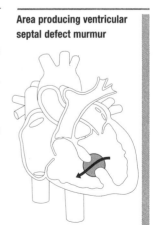

Auscultatory area for ventricular septal defect murmur

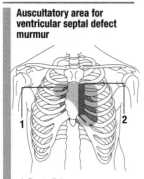

1. Precordial area
2. Left sternal border

The murmur is holosystolic. S_2 is widely split because the aortic valve closes early. If the VSD is large, pulmonary hypertension eventually develops, the S_2 narrows, and the P_2 becomes louder than the A_2. The intensity and pitch vary with the size of the VSD. If pulmonary hypertension becomes severe, shunting may stop and the P_2 and A_2 may merge.

◀◀◀ POSTTEST

1. Describe the sound production of systolic murmurs.
2. How is the sound of an innocent SEM produced?
3. Describe the characteristics of a SEM.
4. How can a SEM be enhanced?
5. What is a ventricular outflow obstruction murmur?
6. Describe the characteristics of a supravalvular pulmonic stenosis murmur.
7. Compare mild pulmonic valvular stenosis with severe pulmonic valvular stenosis.
8. Describe the characteristics of a pulmonic valvular stenosis murmur.
9. Describe the characteristics of a subvalvular pulmonic stenosis murmur.
10. Describe the characteristics of a supravalvular aortic stenosis murmur.
11. Describe the characteristics of an aortic valvular stenosis murmur.
12. What causes subvalvular aortic stenosis?
13. Describe the characteristics of a subvalvular aortic stenosis murmur.
14. How are systolic regurgitation murmurs produced?
15. Describe the characteristics of a tricuspid regurgitation murmur.
16. How can the tricuspid regurgitation murmur be enhanced?
17. How is the sound of a mitral regurgitation murmur produced?
18. Describe the characteristics of a holosystolic mitral regurgitation murmur.
19. Describe the characteristics of an acute mitral regurgitation murmur.
20. Describe the characteristics of a mitral valve prolapse murmur.
21. Which maneuvers enhance the mitral valve prolapse murmur?

Diastolic murmurs

9

PRETEST ▶▶▶

1. How is the sound of a diastolic murmur produced?
2. List the aortic regurgitation murmurs.
3. List the pulmonic regurgitation murmurs.
4. What are the characteristics of a mitral stenosis murmur?
5. What are the characteristics of a tricuspid stenosis murmur?

Anatomy and physiology

At the end of ventricular systole and the beginning of diastole, the aortic and pulmonic valves close. During this time, the A_2-P_2 interval can usually be auscultated. After a brief period of isovolumic relaxation, the mitral and tricuspid valves open, and blood flows from the atria into the ventricles. During this early filling period, a third heart sound (S_3) can sometimes be heard in healthy individuals younger than age 20. Late in diastole, the atria contract, increasing blood flow into the ventricles. Occasionally, a fourth heart sound can be heard during this late filling period. Except for these brief heart sounds, diastole is normally silent. When diastolic murmurs occur, they are heard between the second and first heart sounds (S_2 and S_1) or between the end of the T wave and the beginning of the QRS complex on the electrocardiogram (ECG) waveform.

The regurgitation of blood through the aortic and pulmonic valves may cause diastolic murmurs. Because both valves close at the beginning of diastole, murmurs produced by dysfunctional aortic and pulmonic valves begin early in diastole, immediately after the affected valve closes.

PCG and ECG showing early diastolic aortic regurgitation murmur

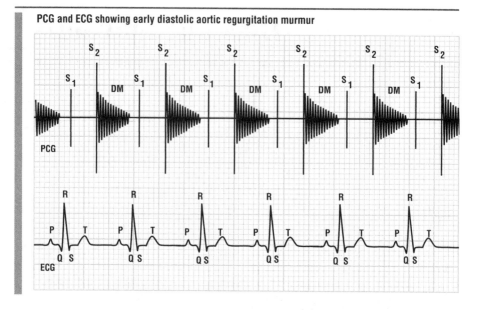

Mitral and tricuspid valve stenosis and conditions that produce turbulent blood flow across normal mitral or tricuspid valves also cause diastolic murmurs. Because these valves open after a period of isovolumic relaxation, these murmurs are heard during middiastole.

Aortic regurgitation murmurs

EARLY DIASTOLIC AORTIC REGURGITATION MURMURS

The early diastolic aortic regurgitation murmur, also called *aortic insufficiency,* can result from rheumatic heart disease, Marfan syndrome, osteogenesis imperfecta, a congenital bicuspid valve, or a dissecting aortic aneurysm. Leakage around a prosthetic aortic valve can also produce this murmur.

Because aortic pressure normally exceeds left ventricular pressure at the beginning of diastole, it forces a retrograde, or backward, flow of blood across an incompetent aortic valve. The turbulent blood flow produces the murmur. In this murmur, A_2 may sound normal, or it may be accentuated if the patient has severe systemic hypertension. (◆1-65) This murmur is commonly associated with a systolic ejection murmur produced by increased left ventricular stroke volume.

Area producing early diastolic aortic regurgitation murmur

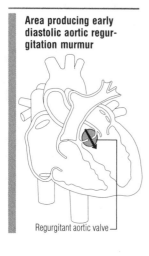

Regurgitant aortic valve

Sound characteristics

The early diastolic aortic regurgitation murmur is usually heard best near the heart's base over the aortic and pulmonic areas, over Erb's point, and near the heart's apex over the mitral area. Because of its usually soft intensity, it's heard best in a quiet environment. The murmur can last throughout most of diastole. It has a high pitch that's heard best using the diaphragm of the stethoscope and a blowing or musical quality. Its timing is diastolic, beginning with A_2. On the ECG waveform, the murmur begins after the T wave and ends just before the QRS complex. It has a decrescendo configuration. (◆1-66)

If aortic regurgitation is associated with aortic root dilation or with a dissecting aneurysm of the ascending aorta, the diastolic murmur may be louder along the right sternal border between the second and fourth intercostal spaces than along the left sternal border between the second and fourth intercostal spaces, respectively. If the patient is elderly or has chronic obstructive pulmonary disease, the murmur may be heard best near the heart's apex over the mitral area. If the murmur is loud, it may be heard over most of the precordium.

Enhancement techniques

This murmur can be enhanced by having the patient sit down, lean forward, and hold his breath after expiration or perform maneuvers that increase aortic diastolic pressure, such as squatting or performing the handgrip exercise.

MIDDIASTOLIC AORTIC REGURGITATION (AUSTIN FLINT) MURMURS

Severe aortic regurgitation may be associated with a middiastolic and presystolic rumbling murmur. This murmur is caused by the aortic regurgitant jet impinging on the normal mitral inflow and augmenting mitral inflow turbulence. The murmur generated by the turbulent blood flow across the mitral valve is called an *Austin Flint murmur;* it may be confused with a murmur of mitral stenosis. (◆1-67)

Sound characteristics

The Austin Flint murmur is usually heard best near the heart's apex over the mitral area. It's introduced with an S_3 rather than an opening snap, as occurs in mitral stenosis. Its intensity is usually soft. It has a low pitch that's heard best using the bell of the stethoscope. The murmur has a rum-

Auscultatory area for early diastolic aortic regurgitation murmur

1. Aortic area
2. Pulmonic area
3. Erb's point
4. Mitral area

Area producing middiastolic aortic regurgitation (Austin Flint) murmur

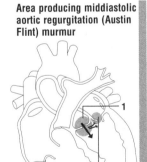

1. Regurgitant aortic valve
2. Anterior mitral leaflet

PCG and ECG showing middiastolic aortic regurgitation (Austin Flint) murmur

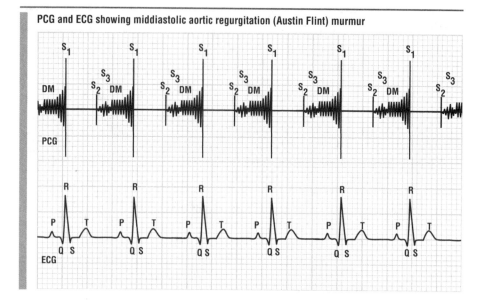

Auscultatory area for middiastolic aortic regurgitation (Austin Flint) murmur

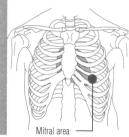

Mitral area

Area producing Graham Steell murmur

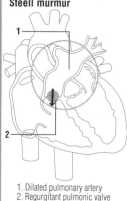

1. Dilated pulmonary artery
2. Regurgitant pulmonic valve

bling quality. Its timing is confined to middiastole or presystole. It's heard just before the QRS complex on the ECG waveform. The murmur has a presystolic crescendo configuration; its middiastolic component has a crescendo-decrescendo configuration. (◆1-68)

Pulmonic regurgitation murmurs

GRAHAM STEELL MURMURS

Because pulmonary artery pressure is normally quite low during diastole, significant regurgitation from a normal pulmonic valve rarely occurs without pulmonary hypertension. Pulmonary hypertension produces pulmonary artery dilation and a relative pulmonic insufficiency because of dilation of the pulmonic valve ring. (◆1-69) The pulmonic regurgitation murmur resulting from pulmonary hypertension is called a *Graham Steell murmur.*

Sound characteristics

The Graham Steell murmur secondary to pulmonary hypertension is usually heard best along the left sternal border over the third and fourth intercostal spaces. It isn't transmitted to the right sternum. Its intensity is usually loud and its duration variable. The murmur has a high pitch that's heard

PCG and ECG showing Graham Steell murmur

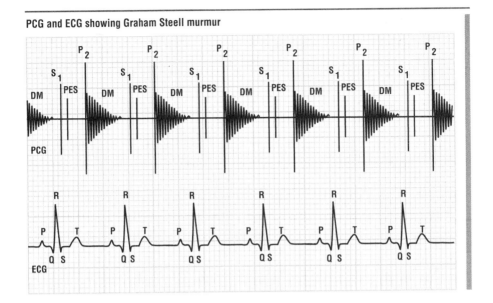

best using the diaphragm of the stethoscope, and it has a blowing quality. Its timing is early diastolic, beginning with a loud P_2. Sometimes an ejection sound can be heard. The murmur is heard after the end of the T wave on the ECG waveform. It has a decrescendo configuration. (◆1-70)

Enhancement techniques
The Graham Steell murmur is intensified during inspiration.

NORMAL PRESSURE PULMONIC VALVE MURMURS
When a murmur results from idiopathic pulmonary artery dilation or congenital pulmonary valve insufficiency, pulmonary artery diastolic pressure isn't elevated. Regurgitation is delayed until right ventricular pressure has fallen below pulmonary artery diastolic pressure during isovolumic relaxation. (◆1-71)

Sound characteristics
The normal pressure pulmonic valve murmur is usually heard best along the left sternal border over the third and fourth intercostal spaces. It isn't transmitted to the right sternum. Its intensity is soft and its duration brief. The murmur has a low pitch that's heard best using the bell of the stethoscope. It has a rumbling quality, and its timing is ear-

Auscultatory area for Graham Steell murmur

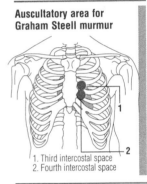

1. Third intercostal space
2. Fourth intercostal space

Area producing normal pressure pulmonic valve murmur

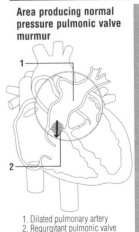

1. Dilated pulmonary artery
2. Regurgitant pulmonic valve

PCG and ECG showing normal pressure pulmonic valve murmur

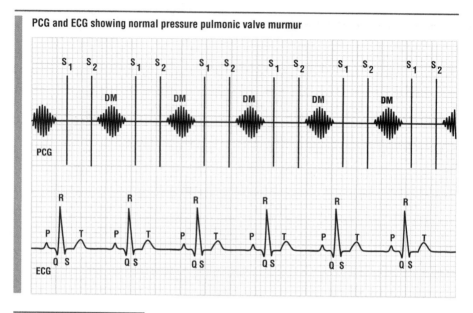

Auscultatory area for normal pressure pulmonic valve murmur

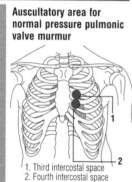

1. Third intercostal space
2. Fourth intercostal space

ly to middiastolic. It begins shortly after P_2 is heard. In the ECG waveform, it begins after the T wave and ends before the P wave. The murmur has a crescendo-decrescendo configuration. (◆1-72)

Enhancement techniques
The normal pressure pulmonic valve murmur is intensified during inspiration.

Other diastolic murmurs

MITRAL STENOSIS MURMURS
Mitral stenosis results from a congenital defect or chronic rheumatic valvulitis. Typically, valvulitis results in scarring and calcification of the valve, causing it to cease normal function. The scarring and calcification create a funnel-shaped opening between the left atrium and left ventricle.

The mitral stenosis murmur starts with an opening snap (OS) after A_2 and isovolumic relaxation (◆1-73); this OS is an important diagnostic feature of mitral stenosis. The murmur is produced by rapid, turbulent blood flow through a rigid, narrowed mitral valve opening. Turbulence increases just before systole; this gives the murmur its characteristic presystolic crescendo.

Area producing mitral stenosis murmur

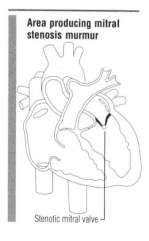

Stenotic mitral valve

PCG and ECG showing mitral stenosis murmur (atrial fibrillation)

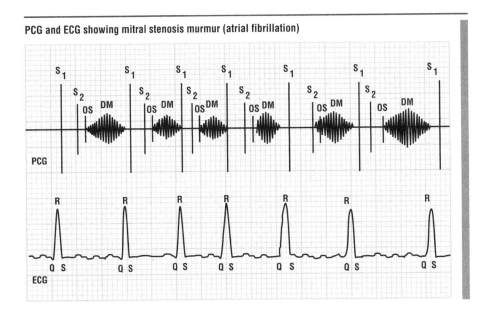

PCG and ECG showing mitral stenosis murmur (normal sinus rhythm)

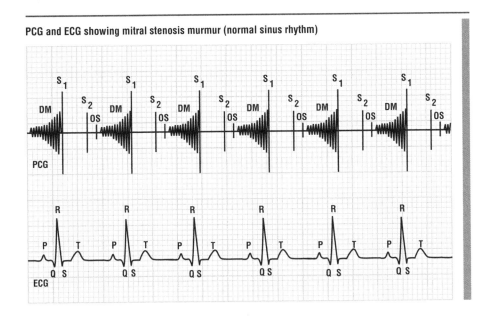

The more severe the stenosis, the longer the duration of the murmur. In severe mitral stenosis, the murmur is holo-diastolic; in moderate stenosis, it appears in early and late diastole. The murmur ends with a loud M_1, which implies that the valve leaflets are mobile; the M_1 is usually palpable. If the valve is calcified, S_1 is soft and an OS isn't heard.

Mitral stenosis is sometimes associated with mitral regurgitation, so a systolic murmur may be heard as well. If mitral stenosis has caused pulmonary hypertension and right-sided heart failure, tricuspid regurgitation may be the dominant murmur heard.

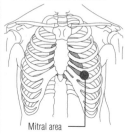

Auscultatory area for mitral stenosis murmur

Mitral area

Sound characteristics

The mitral stenosis murmur is usually heard best near the heart's apex over the mitral area. Its intensity and duration vary. It has a low pitch that's heard best using the bell of the stethoscope and a rumbling quality that has been compared to thunder. The timing of a mitral stenosis murmur is diastolic: It begins after A_2 with an OS and ends with a loud M_1. (◆1-74) However, stenosis severity affects duration of the murmur. On the ECG waveform, the murmur begins just after the T wave and ends during the QRS complex. It has a crescendo-decrescendo configuration. In patients with normal sinus rhythm, the presystolic component has a crescendo configuration. (◆1-75) (See *PCG and ECG showing mitral stenosis murmur [atrial fibrillation]* and *PCG and ECG showing mitral stenosis murmur [normal sinus rhythm],* page 87.)

Enhancement techniques

The mitral stenosis murmur is heard best with the patient in a partial left lateral recumbent position. It can be intensified by maneuvers that increase cardiac output, such as having the patient exercise for a few minutes, raise his legs from a recumbent position, or cough several times.

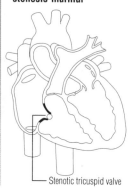

Area producing tricuspid stenosis murmur

Stenotic tricuspid valve

TRICUSPID STENOSIS MURMURS

Tricuspid stenosis rarely occurs as an isolated defect; it's almost always associated with mitral valve defects. The tricuspid stenosis murmur begins shortly after a normal S_2 and is introduced by an OS that can be seen on a phonocardiogram but that is seldom heard. (◆1-76)

Sound characteristics

The tricuspid stenosis murmur is usually heard best over the tricuspid area. Its usually soft intensity increases during inspiration and fades or disappears during expiration.

PCG and ECG showing tricuspid stenosis murmur

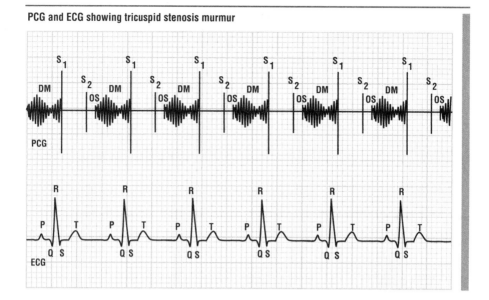

The duration of the murmur varies. It has a low pitch that's heard best using the bell of the stethoscope while the patient is in the partial left lateral recumbent position. The murmur has a rumbling quality. Its timing is mid- to late diastolic: It ends just before S_1. An OS may be heard. On the ECG waveform, the murmur begins just after the T wave and ends just before the QRS complex. In patients with normal sinus rhythm, it has a late diastolic crescendo or crescendo-decrescendo configuration. (◆1-77)

AUSCULTATION TIP *The tricuspid stenosis murmur can be differentiated from the mitral stenosis murmur because the murmur intensifies with inspiration, whereas the mitral stenosis murmur doesn't.*

Auscultatory area for tricuspid stenosis murmur

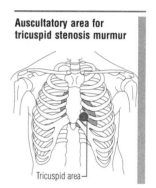

Tricuspid area

◀◀◀ POSTTEST

1. How is the sound of a diastolic murmur produced?
2. Describe the characteristics of an early diastolic aortic regurgitation murmur.
3. How can an early diastolic aortic regurgitation murmur be enhanced?
4. Describe the characteristics of an Austin Flint murmur.
5. Describe the characteristics of a Graham Steell murmur.
6. How is a Graham Steell murmur enhanced?

7. Describe the characteristics of a normal pressure pulmonic valve murmur.
8. How is a normal pressure pulmonic valve murmur enhanced?
9. How is the sound of a mitral stenosis murmur produced?
10. Describe the characteristics of a mitral stenosis murmur.
11. How is a mitral stenosis murmur enhanced?
12. Describe the characteristics of a tricuspid stenosis murmur.

Continuous murmurs

10

PRETEST ▶▶▶

1. How is the sound of a continuous murmur produced?
2. What is a cervical venous hum?
3. Why do the sounds of a patent ductus arteriosus (PDA) murmur change with age?
4. How can you differentiate between a PDA murmur and cervical venous hum murmur?

Two causes

Continuous murmurs are generated by rapid blood flow through arteries or veins or by shunting. Shunting occurs when an abnormal communication is created between the high-pressure arterial system and the low-pressure venous system. The murmurs begin in systole and persist, without interruption, through the second heart sound (S_2) into diastole. Because these murmurs end late in diastole, they're continuous throughout the cardiac cycle.

Cervical venous hum murmurs

The most common continuous murmur is the normal cervical venous hum, caused by rapid downward blood flow through the jugular veins in the lower part of the neck. (◆1-78) The cervical venous hum is present in most individuals, but it's more pronounced in patients with hyperkinetic circulatory states, such as anemia, pregnancy, and thyrotoxicosis. The hum will disappear if the patient performs Valsalva's maneuver, lies down, or has pressure applied over the jugular vein.

Area producing cervical venous hum murmur

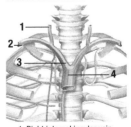

1. Right internal jugular vein
2. Right subclavian vein
3. Right brachiocephalic vein
4. Superior vena cava

91

PCG and ECG showing cervical venous hum murmur

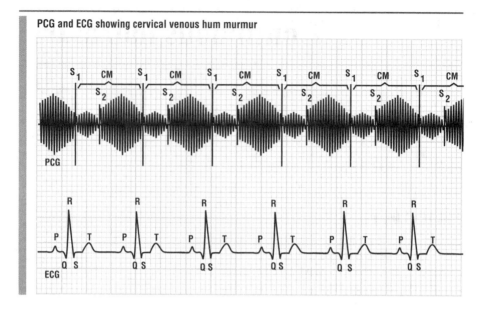

Sound characteristics

The cervical venous hum murmur is heard best over the right supraclavicular fossa when the patient is sitting and his head is turned to the left. It has a faint intensity that increases during diastole. The murmur has a long duration: It lasts throughout the cardiac cycle. Its high pitch is heard best using the diaphragm of the stethoscope. Its quality is soft and its timing continuous. The hum is louder between S_2 and the first heart sound (S_1). (◆1-79) The murmur has a plateau shape in systole and a crescendo-decrescendo configuration in diastole.

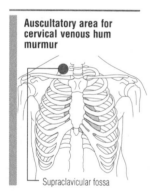

Auscultatory area for cervical venous hum murmur

Supraclavicular fossa

Patent ductus arteriosus murmurs

The patent ductus arteriosus (PDA) murmur is the prototype of continuous murmurs. (◆1-80) Normally, the ductus is open during fetal life and closes shortly after birth. When the ductus remains open, abnormal blood flow, or shunting, occurs from the high-pressure aorta to the low-pressure pulmonary artery. The PDA murmur extends into diastole. A thrill may be palpable in a loud PDA murmur.

PCG and ECG showing patent ductus arteriosus murmur

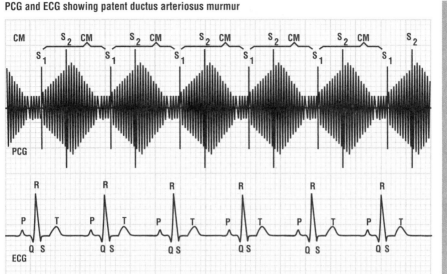

Sound characteristics

The PDA murmur is heard best over the entire left first and second intercostal spaces. Its intensity is faint when the murmur is limited to this area. If the murmur is loud, the systolic component may be heard along the left sternal border or over the mitral area. A loud PDA murmur may radiate to the back between the scapulae. The PDA murmur has a long duration. Its intensity varies, roughly correlating with the ductus size; typically, the murmur reaches maximum intensity late in systole and then fades during diastole.

The PDA murmur has a high pitch heard equally well using the bell or diaphragm of the stethoscope. It has a rough, machinery-like quality, and its timing is continuous. It begins with — or shortly after — a normal S_1 and disappears just before the next S_1. Because the murmur peaks in late systole, S_2 may be difficult to hear. S_2 may be paradoxically split if left ventricular ejection time is prolonged. A third heart sound may be heard over the mitral area. **(♦1-81)** The murmur has a crescendo-decrescendo configuration.

Enhancement techniques

Exercise can increase the intensity and duration of a PDA murmur.

Area producing patent ductus arteriosus murmur

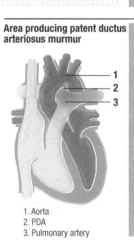

1. Aorta
2. PDA
3. Pulmonary artery

Auscultatory area for patent ductus arteriosus murmur

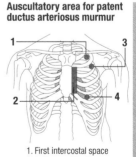

1. First intercostal space
2. Left sternal border
3. Second intercostal space
4. Mitral area

DIFFERENTIATING PDA MURMURS FROM CERVICAL VENOUS HUM MURMURS

To differentiate these PDA murmurs from cervical venous hum murmurs, remember that the hum is loudest above the clavicle, is usually heard better on the right, and can be obliterated by pressing on the jugular vein or placing the patient in a supine position. The venous hum is truly continuous and is usually louder during diastole.

◀◀◀ POSTTEST

1. How is the sound of a continuous murmur produced?
2. What causes a cervical venous hum?
3. Describe the characteristics of a cervical venous hum murmur.
4. Describe the characteristics of a PDA murmur.
5. How can a PDA murmur be enhanced?
6. How can you differentiate between a PDA murmur and a cervical venous hum murmur?

Other auscultatory sounds

11

PRETEST ▶▶▶

1. What are the four types of prosthetic valves?
2. What are the similarities and differences of the sounds produced by the four types of aortic prosthetic valves?
3. How do you know if an aortic prosthetic valve is malfunctioning?
4. What are the similarities and differences of the sounds produced by the four types of mitral prosthetic valves?
5. How do you know if a mitral prosthetic valve is malfunctioning?
6. What produces a pericardial friction rub?
7. What is mediastinal crunch?

Prosthetic valve sounds and murmurs

Prosthetic heart valves — such as the ball-in-cage, tilting-disk, bileaflet, and porcine valves — are used to replace malfunctioning human valves. The aortic and mitral valves are the ones most commonly replaced. The tricuspid valve is replaced occasionally; the pulmonic valve, rarely. The following discussion focuses on the aortic and mitral valves.

AGE ISSUE *Because mechanical valves last a long time, they're usually inserted in children and adults younger than age 40. However, these valves shouldn't be inserted in women of childbearing age because mechanical valve implantation requires the patient to take an oral anticoagulant, such as warfarin, which can cause birth defects if taken during the first trimester of pregnancy.*

Auscultatory area for aortic and mitral prosthetic valve murmurs with systolic ejection murmur

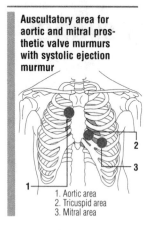

1. Aortic area
2. Tricuspid area
3. Mitral area

Auscultatory area for mitral prosthetic valve murmur with early diastolic murmur

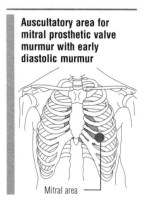

Mitral area

GENERAL CHARACTERISTICS

When a prosthetic valve is surgically placed in the aortic orifice, it's positioned so that it opens during systole and closes at the beginning of diastole. However, when a prosthetic valve is surgically placed in the mitral orifice, it's positioned so that it closes at the beginning of systole and opens during diastole.

An aortic prosthetic valve won't change a normal first heart sound (S_1), and a mitral prosthetic valve will not change a normal second heart sound (S_2). However, prosthetic valves do produce characteristic auscultatory sounds and murmurs. For example, some prostheses produce opening clicks, and all prostheses, regardless of type or position, produce closing clicks. The ball-in-cage valve produces the loudest sounds.

Typically, any disappearance or muffling of previously heard prosthetic valve sounds or murmurs, or the development of a new murmur in a patient with a prosthetic valve, indicates possible prosthetic valve dysfunction.

Almost all prosthetic valves in the aortic position are associated with a soft early systolic or midsystolic ejection murmur. This murmur is heard best over the tricuspid, mitral, and aortic areas and is similar to the mild aortic stenosis murmur.

Some prosthetic valves in the mitral position produce a short, soft, rumbling early diastolic murmur, which is very similar to the mitral stenosis murmur except that it's brief and a presystolic component isn't heard. This murmur is heard best near the heart's apex over the mitral area while the patient is in the partial left lateral recumbent position.

Aortic prosthetic valves

Distinctive sounds and murmurs are associated with the various types of prostheses used in aortic valve replacement. (◆1-82)

AORTIC BALL-IN-CAGE VALVE

A normally functioning aortic ball-in-cage valve prosthesis doesn't change the normal S_1 characteristics. S_1 is followed by an aortic opening click that's louder than S_1; it's sharp and high pitched, and it has a crisp quality. An aortic closing click replaces A_2. Although the aortic ball-in-cage valve

PCG showing aortic ball-in-cage valve murmur

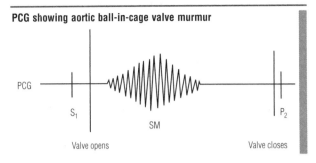

PCG

S_1

SM

P_2

Valve opens   Valve closes

Ball-in-cage valve

Starr-Edwards Silastic Ball Valve. (Photo courtesy of Edwards Lifesciences LLC, Irvine, California. Edwards Lifesciences and Starr-Edwards are trademarks of Edwards Lifesciences Corporation and are registered in the U.S. Patent and Trademark office.)

prosthesis is no longer inserted, patients with this type of prosthesis may still be seen.

Sound characteristics
The murmur generated by an aortic ball-in-cage valve prosthesis is usually heard best near the apex over the mitral and aortic areas and along the left sternal border. Its intensity is usually loud and easy to hear, and its duration variable. It has a medium pitch that's heard best using the diaphragm of the stethoscope. The murmur has a crunchy, harsh quality. Its timing is midsystolic. The interval between the aortic closing click and P_2 is similar to the normal A_2-P_2 interval, and it normally widens during inspiration. The murmur has a crescendo-decrescendo configuration.

Detecting a malfunctioning aortic ball-in-cage valve
Dysfunction of an aortic ball-in-cage valve prosthesis commonly causes the aortic opening click to become soft or absent; the aortic closing click may also be absent. A diastolic murmur and a long systolic ejection murmur (SEM) may appear.

AORTIC TILTING-DISK VALVE
S_1 is unchanged with a normally functioning aortic tilting-disk valve prosthesis. An aortic opening click may be heard, and the interval between the aortic closing click and P_2 is normal. The aortic closing click isn't as loud as the click heard with a ball-in-cage prosthesis.

Sound characteristics
The murmur generated by an aortic tilting-disk valve prosthesis is usually heard best near the apex over the mitral and aortic areas and along the left sternal border. Its intensity is

Tilting-disk valve

Medtronic Hall Mitral Valve. (Photo courtesy of Medtronic, Inc.)

PCG showing aortic tilting-disk valve murmur

PCG

S_1

SM

P_2

Valve opens

Valve closes

Dotted line indicates that valve opening may or may not be heard.

usually soft (grade ⅔) and its duration short. It has a medium pitch heard best with the diaphragm of the stethoscope. The murmur has a rough or harsh quality. Its timing is systolic. The interval between the aortic closing click and P_2 is similar to the normal A_2-P_2 interval, and it widens during inspiration. The murmur has a crescendo-decrescendo configuration.

Detecting a malfunctioning aortic tilting-disk valve

Dysfunction of an aortic tilting-disk valve prosthesis commonly causes some or all of the following changes: the aortic closing click may be absent, a diastolic murmur may be heard, or a longer SEM may appear.

AORTIC BILEAFLET VALVE

S_1 is normal with a normally functioning aortic bileaflet valve prosthesis. The aortic opening click is commonly inaudible, but the closing click is loud and distinct.

Sound characteristics

The murmur generated by an aortic bileaflet valve prosthesis is usually heard best near the apex over the mitral and

Bileaflet valve

SJM Regent Valve. (Photo copyright St. Jude Medical, Inc. 2005. This image is provided courtesy of St. Jude Medical, Inc. All rights reserved. SJM Regent is a registered trademark of St. Jude Medical, Inc.)

PCG showing aortic bileaflet valve murmur

PCG

S_1

SM

P_2

Valve opens

Valve closes

PCG showing aortic porcine valve murmur

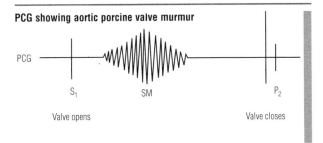

Porcine valve

Carpentier-Edwards Duraflex Low Pressure Porcine Mitral Biopros-thesis Valve. (Photo courtesy of Edwards Lifesciences LLC, Irvine, California. Edwards Lifesciences and Starr-Edwards are trademarks of Edwards Lifesciences Corporation and are registered in the U.S. Patent and Trademark office.)

aortic areas and along the left sternal border. Its intensity is usually soft (grade $\frac{2}{6}$) and its duration short. It has a medium pitch that's heard best using the diaphragm of the stethoscope. The murmur has a rough or harsh quality, and its timing is systolic. The interval between the aortic closing click and P_2 is similar to the normal A_2-P_2 interval, and it widens during inspiration. The murmur has a crescendo-decrescendo configuration.

Detecting a malfunctioning aortic bileaflet valve
Dysfunction of the aortic bileaflet valve prosthesis may cause the aortic closing click to disappear, a diastolic murmur to appear, or both. A longer SEM also may appear.

AORTIC PORCINE VALVE
S_2 is normal with a properly functioning aortic porcine valve prosthesis. There's no opening click in a properly functioning porcine valve.

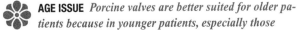 **AGE ISSUE** *Porcine valves are better suited for older patients because in younger patients, especially those younger than age 40, they're prone to calcification. Furthermore, porcine valves gradually deteriorate and must be replaced, making them less desirable for a patient who wants to avoid future surgery. However, because these valves don't require anticoagulation therapy, they may be an option for women who are pregnant or of childbearing age.*

Sound characteristics
The murmur generated by an aortic porcine valve prosthesis is usually heard best near the apex over the mitral and aortic areas and along the left sternal border. Its intensity is usually soft (grade $\frac{2}{6}$) and its duration short. It has a medium pitch that's heard best using the diaphragm of the stethoscope. The murmur has a rough or harsh quality, and its timing is systolic. The interval between the aortic clos-

ing sound and P_2 is similar to the normal A_2-P_2 interval, and it widens during inspiration. The murmur has a crescendo-decrescendo configuration.

Detecting a malfunctioning aortic porcine valve
Dysfunction of an aortic porcine valve prosthesis may cause a diastolic murmur to appear. A longer SEM also may appear.

Mitral prosthetic valves

Distinctive sounds and murmurs are associated with the various types of prostheses used in mitral valve replacement. (♦1-83)

MITRAL BALL-IN-CAGE VALVE
The mitral closing click replaces M_1 with a normally functioning mitral ball-in-cage valve prosthesis. Patients with this type of prosthetic valve have an opening click after S_2 and may have middiastolic clicks or short, soft (grade ⅖), rumbling diastolic murmurs that are heard best near the apex over the mitral area with the patient in a partial left lateral recumbent position.

Sound characteristics
M_1 is loud and higher in frequency. A mitral opening click follows a normal S_2.

Detecting a malfunctioning mitral ball-in-cage valve
Dysfunction of a mitral ball-in-cage valve prosthesis may cause some or all of the following changes: the sounds normally associated with the valve may vary in intensity in a patient with normal sinus rhythm; a short interval between

---

PCG showing mitral ball-in-cage valve sounds

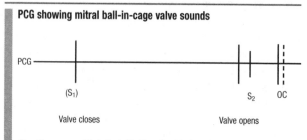

Parentheses around S_1 indicate that S_1 will not be heard because valve closure is louder. Dotted line indicates opening click (OC) may or may not be heard.

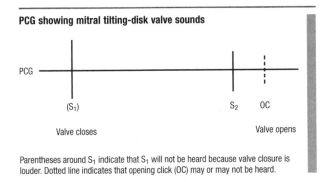

PCG showing mitral tilting-disk valve sounds

PCG

(S_1) S_2 OC

Valve closes   Valve opens

Parentheses around S_1 indicate that S_1 will not be heard because valve closure is louder. Dotted line indicates that opening click (OC) may or may not be heard.

A_2 and the mitral opening click may occur, indicating high left atrial pressure; a holosystolic murmur may occur, indicating mitral regurgitation; or the intensity and duration of the diastolic murmur may change, indicating mitral obstruction.

MITRAL TILTING-DISK VALVE
A normally functioning mitral tilting-disk valve prosthesis replaces M_1 with a mitral closing click. This closing click is always distinctly audible and high pitched; its maximum intensity is located near the apex over the mitral area.

Sound characteristics
A mitral opening click may follow a normal S_2.

Detecting a malfunctioning mitral tilting-disk valve
Dysfunction of a mitral tilting-disk valve prosthesis may cause some or all of the following changes: the mitral closing click may be absent, a new diastolic murmur may develop, a previously auscultated diastolic murmur may intensify, or a holosystolic mitral regurgitation murmur may appear.

MITRAL BILEAFLET VALVE
A normally functioning mitral bileaflet valve prosthesis replaces M_1 with a mitral closing click that's loud and high pitched. The mitral opening click, however, may be inaudible. If it's audible, it follows a normal S_2. In some patients with a mitral bileaflet valve prosthesis, a short middiastolic rumble occurs; this rumble is heard best near the apex over the mitral area.

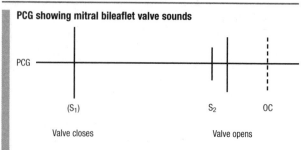

PCG showing mitral bileaflet valve sounds

PCG

(S_1) S_2 OC

Valve closes   Valve opens

Parentheses around S_1 indicate that S_1 will not be heard because valve closure is louder. Dotted line indicates that opening click (OC) may or may not be heard.

Sound characteristics
The first sound is replaced by a higher-pitched sound. The mitral opening click, though rarely heard, follows a normal S_2.

Detecting a malfunctioning mitral bileaflet valve
Dysfunction of a mitral bileaflet valve prosthesis may cause a holosystolic murmur, a new diastolic murmur, or both to appear.

MITRAL PORCINE VALVE
A normally functioning mitral porcine valve prosthesis generates a mitral closing sound that's heard over the mitral area and sounds like a normal M_1. No opening sound is heard with this type of valve.

Sound characteristics
This sound is heard best near the apex over the mitral area.

Detecting a malfunctioning mitral porcine valve
Dysfunction of a mitral porcine valve prosthesis may cause a holosystolic mitral regurgitation murmur, a diastolic rumble associated with mitral stenosis, or both to appear.

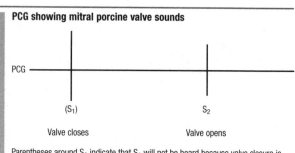

PCG showing mitral porcine valve sounds

PCG

(S_1) S_2

Valve closes   Valve opens

Parentheses around S_1 indicate that S_1 will not be heard because valve closure is louder.

Pericardial friction rubs

When inflamed pericardial surfaces rub together, they produce characteristic high-pitched friction noises known as *pericardial friction rubs.* (◆1-84) These rubs are a classic sign of pericarditis (inflammation of the pericardium, the sac surrounding the heart), which can be caused by viral or bacterial infections, radiation therapy to the chest, or cardiac trauma. Pericardial friction rubs may also be heard after a pericardiotomy and for a few hours to a few days after a myocardial infarction.

Sound characteristics
A pericardial friction rub is usually heard best, and is sometimes palpable, over the tricuspid and xiphoid areas. It's usually loud and may get louder during inspiration. The rub has a high pitch heard best with the diaphragm of the stethoscope, and it has a grating or scratchy quality. It's usually heard during each cardiac cycle. It has a systolic component and an early and late diastolic component. (◆1-85) The diastolic components may last for only a few hours.

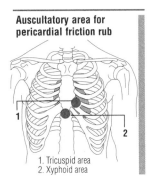

Auscultatory area for pericardial friction rub

1. Tricuspid area
2. Xyphoid area

🔊 **AUSCULTATION TIP** *To differentiate a pericardial friction rub from a pleural friction rub, have the patient hold his breath. When he does so, a pericardial friction rub will persist, but a pleural friction rub will become inaudible.*

Mediastinal crunch

Heart movements can displace air that's present in the mediastinum; this displacement produces crunchy noises, known as *mediastinal crunch,* which may occur randomly or in a consistent pattern. (◆1-86) Patients with mediastinal crunch commonly have subcutaneous emphysema (air trapped beneath the skin). You can assess for this condition by palpating for crepitation in the neck.

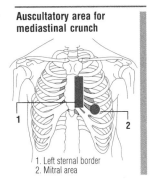

Auscultatory area for mediastinal crunch

1. Left sternal border
2. Mitral area

Sound characteristics
The noises produced by mediastinal crunch are usually heard best along the left sternal border with the patient in a sitting position. The noises have a crunching quality, which may become louder during inspiration. (◆1-87)

◀◀◀ POSTTEST

1. Which two valves are most commonly replaced?
2. What are the four types of prosthetic valves?
3. Describe the characteristics of an aortic ball-in-cage valve murmur.
4. Which changes in sounds can you expect if an aortic ball-in-cage valve malfunctions?
5. Describe the characteristics of an aortic tilting-disk murmur.
6. Which changes in sounds can you expect if an aortic tilting-disk valve malfunctions?
7. Describe the characteristics of an aortic bileaflet valve murmur.
8. Which changes in sounds can you expect if an aortic bileaflet valve malfunctions?
9. Describe the characteristics of an aortic porcine valve murmur.
10. Which changes in sounds can you expect if an aortic porcine valve malfunctions?
11. Describe the characteristics of a mitral ball-in-cage valve murmur.
12. Which changes in sounds can you expect if a mitral ball-in-cage valve malfunctions?
13. Describe the characteristics of a mitral tilting-disk murmur.
14. Which changes in sounds can you expect if a mitral tilting-disk valve malfunctions?
15. Describe the characteristics of a mitral bileaflet valve murmur.
16. Which changes in sounds can you expect if a mitral bileaflet valve malfunctions?
17. Describe the characteristics of a mitral porcine valve murmur.
18. Which changes in sounds can you expect if a mitral porcine valve malfunctions?
19. What causes a pericardial friction rub?
20. Describe the characteristics of a pericardial friction rub.
21. How can you differentiate between a pericardial and a pleural friction rub?
22. What is mediastinal crunch?
23. Describe the characteristics of mediastinal crunch.

FOUR
Breath sound fundamentals

The respiratory system and auscultation

12

PRETEST ▶▶▶

1. What is the primary function of the respiratory system?
2. Name two anatomic structures of the upper airway.
3. Name the anatomic structures of the lungs that are responsible for oxygen–carbon dioxide exchange.
4. What are the alveoli?
5. What are the roles of the mucus layer and the cilia in the airway?
6. Where are the apex and base of the lung located?
7. What is surfactant?
8. Name the two blood supplies in the lungs.
9. How is the lower airway innervated?
10. What are the pleurae?
11. What are the muscles of respiration?
12. Describe the process of normal breathing.
13. Are ventilation and blood flow equally distributed in the lungs?
14. Which three chest wall surfaces are included in breath sound auscultation?
15. Is the diaphragm of a stethoscope ideally suited to transmit high-pitched or low-pitched sounds?
16. What is the auscultatory sequence?
17. What does *contralateral comparison* mean?
18. Does patient position affect breath sounds?

Anatomy and physiology

The primary function of the respiratory system is the exchange of oxygen (O_2) and carbon dioxide (CO_2) between the alveoli and the pulmonary circulation.

During inspiration, air enters the upper airway and travels through the lower airways until it reaches the alveoli.

$CO_2 - O_2$ exchange and airflow pattern during inspiration

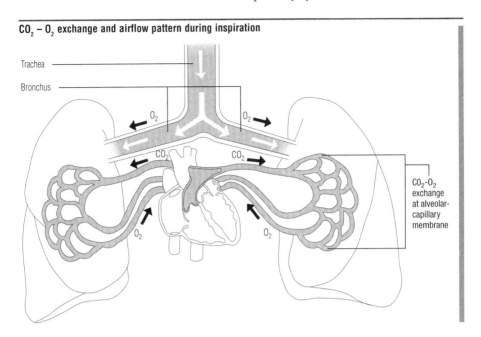

Trachea

Bronchus

CO_2-O_2 exchange at alveolar-capillary membrane

Each alveolus is surrounded by multiple capillaries. During systole, deoxygenated blood returning from the body's cells is pumped from the right ventricle through the arterial pulmonary circulation to the alveolar capillaries. CO_2 diffuses from the capillary blood across the alveolar capillary membrane and enters the alveolar air. Simultaneously, O_2 from the inspired atmospheric air in the alveolus crosses the alveolar capillary membrane and enters the pulmonary capillary blood.

During expiration, CO_2 is exhaled from the lungs. Oxygenated capillary blood travels to the left side of the heart and is pumped from the ventricle into arterial circulation to the cells of the body, where cellular respiration occurs. (See *CO_2-O_2 exchange and airflow pattern during expiration,* page 108.)

UPPER AIRWAY

The upper airway consists of the nasal cavities and the pharynx. Air enters the body through the nostrils and passes through the nasal cavities. Large hairs located in the nostrils filter out foreign particles. The lateral walls of the nasal cavities curve, forming turbinates (conchae), which increase the mucosal surface area and produce turbulent airflow.

Location of nasal cavity and pharynx

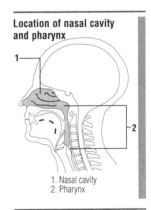

1. Nasal cavity
2. Pharynx

Location of turbinates and airflow pattern in the nasal cavity

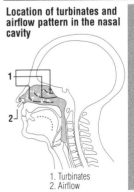

1. Turbinates
2. Airflow

$CO_2 - O_2$ exchange and airflow pattern during expiration

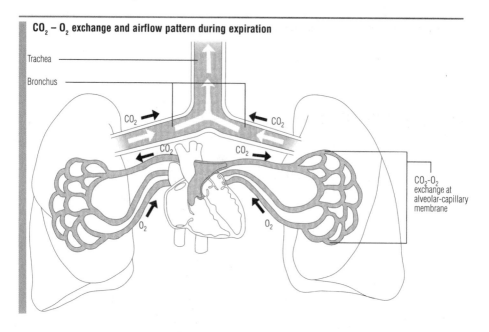

Trachea

Bronchus

CO_2

CO_2

CO_2

CO_2

O_2

O_2

CO_2-O_2 exchange at alveolar-capillary membrane

Location of nasopharynx, oropharynx, and hypopharynx

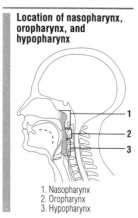

1. Nasopharynx
2. Oropharynx
3. Hypopharynx

The lining of the nasal cavities contains a dense capillary network and serous glands that warm and humidify the inspired air. Seromucous glands produce a sticky film over the lining that traps particles against it during airflow.

The pharynx is divided into three sections: the nasopharynx, the oropharynx, and the hypopharynx. The nasopharynx begins at the posterior openings of the nasal cavities and extends to the soft palate; the oropharynx, the middle section, is the common passageway for food and air; and the hypopharynx, the lower section, extends from the roof of the tongue to the glottic area.

AGE ISSUE *In children, the larynx and glottis are positioned higher in the neck, creating a greater risk for aspiration.*

LOWER AIRWAY

The lower airway is called the *tracheobronchial tree*. It begins at the level of the cricoid cartilage of the larynx and ends at the distal bronchioles, which open into the alveoli.

The larynx, which contains the epiglottis and vocal cords, has a role in ventilation and phonation and helps protect the lower airway from aspiration.

Location of trachea, right and left mainstem bronchi, and segmental and lobar bronchi

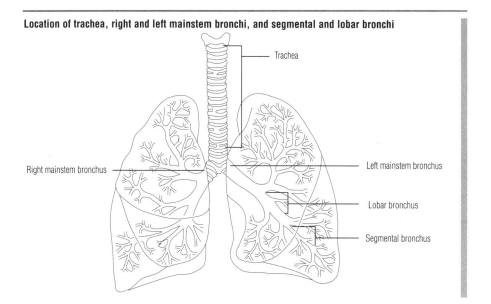

Trachea

The adult trachea is about 1" (2.5 cm) in diameter and 4" (10 cm) long to the bifurcation, which is also called the *carina.* Anteriorly, the trachea begins just below the larynx and is located within the thoracic cavity, beneath the upper two-thirds of the sternum. Posteriorly, it begins at the level of the sixth cervical or first thoracic vertebra and extends to about the fifth thoracic vertebra.

 AGE ISSUE *In children, the trachea is shorter with a more acute angle at the bifurcation of the right bronchus. The smaller diameter of the trachea results in increased airway resistance.*

Mainstem bronchus

The trachea divides into a right and left mainstem bronchus. The left mainstem bronchus leaves the trachea at a sharper angle than the right mainstem bronchus and passes under the aortic arch before entering the lung. The mainstem bronchi enter the lungs at the hila, where the lung tissue attaches to the mediastinum. The bronchi course downward and immediately divide into lobar bronchi, which subdivide into segmental bronchi.

The airways continue to systematically branch, narrow, shorten, and increase in number toward the lung periphery.

Location of larynx, epiglottis, and vocal cords

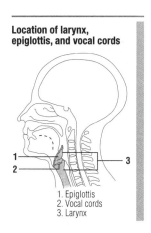

1. Epiglottis
2. Vocal cords
3. Larynx

Location of terminal bronchioles and alveoli

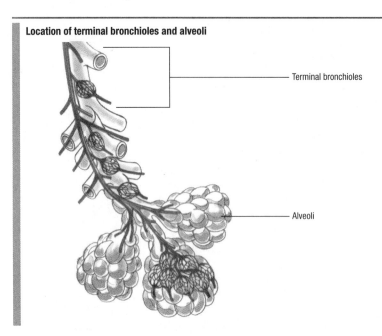

Terminal bronchioles

Alveoli

They divide about 25 times before ultimately opening into the terminal, or respiratory, bronchioles.

Terminal bronchioles

The terminal bronchioles, which have a few alveoli budding from their walls, branch into alveolar ducts that are completely lined with alveoli. These airways and their alveoli form the respiratory lobules of the lungs. The terminal bronchioles account for the last six or seven airway divisions. The distance from the terminal bronchiole to the most distal alveolus is only about 5 mm; however, most of the lung surface area available for gas exchange is contained in these spaces.

Conducting airways

The airways between the nose and the terminal bronchioles are called the *conducting airways.* About 150 cc of inspired air remains in the conducting airways during each breath. This air, called *anatomic dead space,* is not involved in gas exchange.

The large airways below the glottis contain cartilage, smooth muscle, elastic fibers, connective tissue, and mucous glands. Anteriorly, the trachea and mainstem bronchi are composed of 20 rings of U-shaped cartilage. The cartilage that opens posteriorly is connected by a band of smooth muscle and fibroelastic tissue. About 30% of the wall thickness in the mainstem and lobar bronchi is cartilage. This cartilage becomes irregular and platelike and eventually disappears in the smaller airways, which consist solely of smooth muscle and elastic fibers.

The mucous and bronchial glands located in the bronchial walls secrete a serous mucus blanket onto the inside airway walls. This secreted mucus forms a thin, sticky layer that lies atop of ciliated columnar epithelial cells, which line nearly all of the respiratory tract between the larynx and the terminal bronchioles. The mucus layer traps foreign particles in the inspired air before they can enter the lungs. Cilia move in a wavelike motion, propelling the sticky mucus layer toward the pharynx where it can be coughed out or swallowed. Together, the mucus layer and cilia protect the lungs, acting as a major defense mechanism against inhaled particles.

Lung lobes

Each lung is divided into lobes. The right lung has three lobes: the upper, middle, and lower. The left lung has only an upper and a lower lobe. The lung's apex (superior aspect) is located near the clavicle, and its base (inferior aspect) is located near the diaphragm. (See *Lung lobes*, page 112.)

ALVEOLI

The terminal bronchioles divide into 2 to 5 alveolar ducts, each of which consists of 10 to 16 alveoli. Alveoli contain three cell types. Type I, the largest, is the lining cell and accounts for 95% of the alveolar surface area. The type II cell produces surfactant, a mixture of phospholipids. The macrophage, the third cell type, acts as a phagocytic defense mechanism against infection.

The adult respiratory system contains about 300 million alveoli. In a healthy adult, the surface area of the alveolar-capillary membrane available for O_2-CO_2 exchange is 70 to 85 m^2, about the size of a tennis court.

AGE ISSUE *Adults have about 10 times the number of alveoli as infants, who have alveoli that are larger with less elastic recoil. The number of alveoli reaches that of an adult by age 12.*

The alveolar surface is covered by a fluid layer containing surfactant. Surfactant mixes with pulmonary fluids and maintains alveolar stability by lowering the alveolar surface tension, thus preventing the air-filled alveoli from collapsing at low volumes from the weight of capillary blood.

Lung lobes

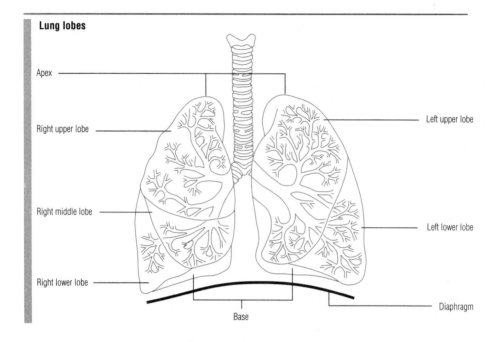

Apex

Right upper lobe

Right middle lobe

Right lower lobe

Left upper lobe

Left lower lobe

Diaphragm

Base

🌸 **AGE ISSUE** *In older adults, elastic recoil in the alveoli decreases and surface area for gas exchange is reduced as alveoli septa are lost.*

LOWER AIRWAY INNERVATION

Two neural pathways, the parasympathetic and sympathetic nervous systems, innervate the tracheobronchial tree. The relative balance of the two neural pathways and their effect on the smooth muscle, which lines the tracheobronchial tree, results in bronchial smooth-muscle tone.

The parasympathetic nervous system in the larynx, trachea, and bronchi includes vagal nerve fibers and irritant receptors mediated by the neurotransmitter acetylcholine. Parasympathetic stimulation precipitates bronchoconstriction, coughing, and mucus discharge from the bronchial glands. Irritant receptors within the airways are believed to be responsible for increasing ventilation and precipitating coughing.

Sympathetic innervation of the lower airways usually arises from the thoracic nerves T1 through T5 and is mediated by the postganglionic hormone transmitter norepinephrine. Sympathetic stimulation of beta$_2$-adrenergic receptors throughout the lower airways facilitates smooth-muscle relaxation, resulting in bronchodilation.

Pulmonary circulation

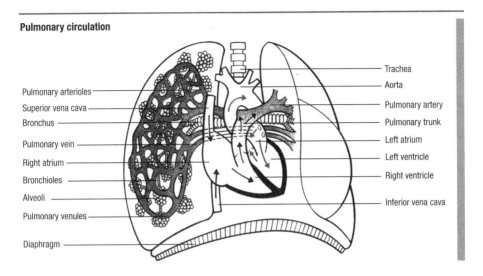

Secretory mast cells on the surface of the large airways also may be stimulated or affected by drugs, hormones, antigens, and other messenger cells to precipitate or control bronchoconstriction.

PULMONARY AND BRONCHIAL CIRCULATION

Two blood supplies, the pulmonary and bronchial circulations, facilitate gas exchange and nourish the airways and other lung tissues.

The pulmonary circulation begins as the pulmonary artery leaves the right ventricle. The pulmonary artery divides into segmental arteries, which subdivide, ending in the pulmonary capillary bed surrounding the alveoli. The capillaries then join to form venules, which converge to form veins, then pulmonary veins. The pulmonary veins end in the left atrium.

The bronchial circulation usually originates in the aorta and branches along the bronchi to provide oxygen and nutrients to the lung tissue and to clear metabolic wastes from the airways and other lung tissues. Oxygenated blood is pumped from the aorta to the tracheobronchial tree and travels through arterioles along the airways. Deoxygenated blood from the bronchial circulation is returned to the left atrium, accounting for the normal 2% to 3% right-to-left shunt.

PLEURAE

The inner surface of the chest wall and the outer surface of the lungs are joined by thin, membranous tissues called the *pleurae*. The parietal pleura covers the inner surface of the chest wall, and the visceral pleura covers the outer surfaces of the lungs. A small amount of fluid exists between the two layers. The pleurae separate into the right and left pleural cavities, which contain the lungs. During expansion and contraction of the chest wall and lungs, the pleurae glide smoothly and silently over each other.

The mechanics of breathing

ANATOMY AND PHYSIOLOGY

The respiratory muscles help the chest cavity expand and contrast. Pressure differences between atmospheric air and the lungs help produce air movement.

Thorax

The *thorax* is the bony structure that protects the vital organs (the heart and lungs) and permits chest expansion during inspiration. Twelve pairs of ribs are attached in a hinge-like manner posteriorly to the vertebral column. This hinge-like attachment allows the chest wall to be mobile during the respiratory cycle. The superior six pairs of ribs are attached anteriorly to the sternum.

Diaphragm

The *diaphragm*, the primary muscle of respiration, is composed of two dome-shaped hemidiaphragms that are attached to the lower edge of the rib cage. The right hemidiaphragm sits higher than the left. The diaphragm forms the inferior "floor" of the thorax. During inspiration, the diaphragm flattens and descends toward the abdomen. During expiration, it relaxes and ascends to its resting, dome-shaped configuration.

External intercostal muscles

The *external intercostal muscles* are also involved in respiration. These muscles, located between the ribs and innervated by the intercostal nerves, contract during inspiration to stabilize, elevate, and expand the rib cage.

Posterior view of thorax

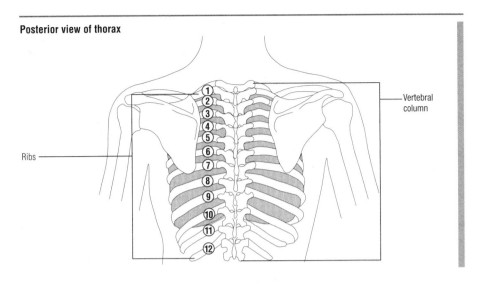

Anterior view of thorax

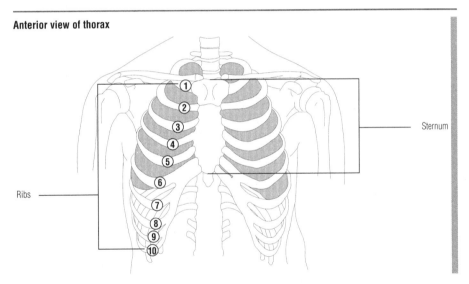

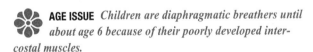

AGE ISSUE *Children are diaphragmatic breathers until about age 6 because of their poorly developed intercostal muscles.*

Accessory muscles of respiration

The *accessory muscles of respiration* — the sternocleidomastoid, scalene, trapezius, and rhomboid muscles — are involved in labored or forceful breathing. The sternocleidomastoid muscle elevates the sternum, increasing the anterior-to-posterior chest diameter. The scalene muscles elevate and fix the first two ribs. The trapezius and rhomboid muscles are activated during respiration but probably have no effect on forceful inspiration. The contraction of external intercostal muscles may also be obvious in extremely labored breathing.

 AGE ISSUE *Age-related anatomic changes that affect the respiratory system include increased anteroposterior diameter of the chest wall (reducing rib mobility and muscle contraction), increased chest wall stiffness with weakened diaphragmatic muscles (increasing use of accessory muscles), and elevated ribs and flattened diaphragm (decreasing chest expansion). Also, because of such factors as osteoporosis and vertebral collapse, kyphosis advances with age, decreasing lung expansion.*

Abdominal muscles

Abdominal muscles usually don't participate actively in relaxed breathing but play an important role during forceful expiratory efforts, such as coughing or sneezing. The internal intercostal muscles also contract forcefully during expiration.

NORMAL BREATHING

During inspiration, the diaphragm and external intercostal muscles are activated. Diaphragmatic contraction flattens the domed diaphragm and expands the lower rib cage, forcing the abdominal contents downward and out, thus increasing the longitudinal lung size. External intercostal muscle contraction stabilizes the rib cage and moves it outward and upward. The posterior cricoarytenoid muscle contracts to open the glottis. As the thorax expands, intrapleural pressures become subatmospheric, resulting in lung expansion and decreased intrapulmonary pressures. Inspired air flows from higher atmospheric pressure into the airways. At the end of inspiration, diaphragmatic movement declines and the air inflow gradually slows.

During expiration, the thorax and the elastic recoil force of the lungs return to their resting positions, increasing intrapleural pressures. This increased pressure forces air to flow out of the lower and upper airways. At the end of expiration, during quiet breathing, all muscles are relaxed and the diaphragm has returned to its resting position.

Diaphragm during inspiration

Diaphragm during expiration

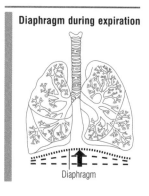

Diaphragm

DISTRIBUTION OF VENTILATION
AND BLOOD FLOW

During normal breathing in the upright position, air movement (ventilation) is greatest in the lung bases or dependent lung regions. Because of the pull of gravity, the alveoli in the lung bases collapse to a smaller size during expiration than do the alveoli in the lung apices. During inspiration, the alveoli in the lung bases open more easily, causing ventilation in the bases to be greater than in the apices.

 AGE ISSUE *In the older adult, airway closure and a loss of surface area results in decreased ventilation at the bases of the lungs.*

Gravity

Blood perfusion in the lungs also depends on gravity; the dependent lung regions receive a larger proportion of the cardiac output than the nondependent regions. When the patient is in the upright position, blood flow patterns in the apices are usually different from those in the bases. In the lung apices, alveolar (atmospheric) pressures exceed capillary hydrostatic pressure, and the capillaries remain partially collapsed during quiet breathing; this results in limited blood flow. In the lung bases, pulmonary arterial and venous hydrostatic pressures exceed atmospheric pressure, and the capillaries remain open throughout quiet breathing; this results in proportionately more blood flow to the lung bases than to the apices.

Posture

Postural changes affect the distribution of air and blood flow throughout the lungs. In the supine position, the apices and bases receive equal amounts of air and blood flow. In the lateral position, dependent lung regions receive more air and blood flow than nondependent regions.

Size and shape

The dimensions and shape of the tracheobronchial tree also affect the airflow pattern. During inspiration, air velocity and the airflow rate decrease as the cross-sectional airway area increases. Consequently, the larger airways conduct the same volume of air as do the more numerous smaller airways. This means that inspired air travels 100 times faster in the trachea than it does in the terminal bronchi.

Beyond the terminal bronchi, the forward movement of air is minimal. In the terminal bronchioles, diffusion is the

Anterior thoracic structures and landmark lines

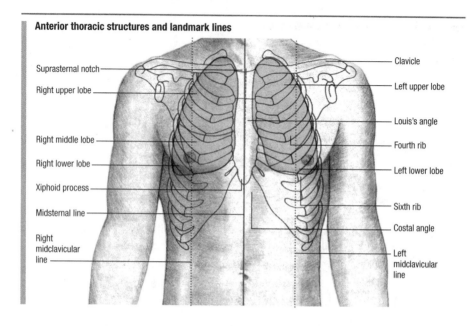

Right lateral landmark lines

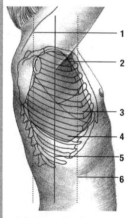

1. Posterior axillary line
2. Right upper lobe
3. Right middle lobe
4. Right lower lobe
5. Midaxillary line
6. Anterior axillary line

primary method of air movement. The mean airflow velocity during inspiration is about 100 m/second in the trachea but less than 1 cm/second in the terminal bronchioles.

Auscultatory areas

THORACIC STRUCTURES AND LANDMARKS

Accurate interpretation of breath sounds depends on a systematic approach to auscultation; it also requires the ability to describe the location of abnormal findings in relation to bony structures and anatomic landmark lines.

Breath sounds are auscultated over the anterior chest wall surface, the lateral chest wall surfaces, and the posterior chest wall surface.

ANTERIOR CHEST WALL SURFACE

The lung apices extend ¼" to 1½" (2 to 4 cm) above the clavicles. The trachea bifurcates at the level of the sternal angle, the junction between the manubrium and the body of the sternum. The ribs and intercostal spaces provide precise horizontal landmarks to describe the location of breath sounds; the second rib and second intercostal space serve as reference points.

Posterior thoracic structures and landmark lines

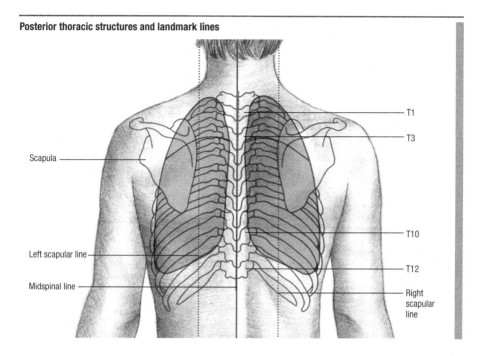

Vertical landmark lines on the anterior chest wall surface include the midclavicular lines and the midsternal line. The right and left midclavicular lines extend downward from the center of each clavicle. The midsternal line bisects the sternum. The right lung base crosses the sixth rib at the right midclavicular line, and the left lung base crosses the seventh rib at the left midclavicular line.

LATERAL CHEST WALL SURFACES

The lateral chest wall surfaces are also divided using landmark lines. The anterior axillary line extends downward from the anterior axillary fold, the midaxillary line extends downward from the apex of the axilla, and the posterior axillary line extends downward from the posterior axillary fold.

POSTERIOR CHEST WALL SURFACE

Bony structures underlying the posterior chest wall surface also provide landmarks to locate breath sounds. The scapulae's inferior borders, located at about the same level as the seventh rib, serve as reference points. The numbered thoracic vertebrae provide horizontal landmarks.

Left lateral landmark lines

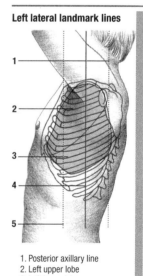

1. Posterior axillary line
2. Left upper lobe
3. Left lower lobe
4. Midaxillary line
5. Anterior axillary line

🔊 **AUSCULTATION TIP** *When auscultating the posterior chest wall, keep in mind that you'll be mostly hearing the lower lobes due to the anatomical positions of the lobes.*

Landmark lines on the posterior chest wall surface provide vertical reference points. The midscapular lines extend downward from the inferior angle of each scapula, and the vertebral line extends downward over the vertebrae.

Auscultatory techniques

USING THE STETHOSCOPE

The stethoscope allows you to hear breath sounds transmitted through the chest wall. Most stethoscopes have a diaphragm and a bell, with one or two tubes leading to the binaural headpiece and earpieces. The diaphragm is used to listen for high-pitched breath sounds; the bell is used to listen for low-pitched breath sounds. (See *Parts of the stethoscope,* page 10.) Applying the stethoscope firmly to the chest wall amplifies high-frequency sounds. However, if too much pressure is applied when using the bell, the stretched skin functions as a diaphragm and filters out low-pitched sounds.

The stethoscope tubing should be no longer than 10" to 12" (25 to 30 cm). It should be tightly attached to the binaural headpiece and the stethoscope body to prevent air leakage, which could result in the loss of sound energy. The earpieces must be securely attached to the binaural headpiece, which removes extraneous noise from the environment, to avoid any sound loss. They should fit tightly and should be placed into the ears in an anterior direction so that they conform to the direction of the ear canals.

🔊 **AUSCULTATION TIP** *Check the stethoscope to ensure that the diaphragm and bell are locked into place before auscultating. Some stethoscopes have a rotating bell or diaphragm that may become disengaged, which would block or muffle breath sounds.*

LISTENING TO BREATH SOUNDS

Breath sounds have a wide range of sound frequencies, many near the lower threshold of human hearing. Consequently, the environment for auscultation should be as quiet as possible so that you can hear breath sounds clearly and distinctly. Close the door to the room and eliminate extraneous noises and conversations; turn off televisions and radios. You'll need to develop good concentration skills so that noises from such devices as I.V. pumps, oxygen delivery systems, and ventilators don't interfere with auscultation.

During auscultation, press the diaphragm of the stethoscope firmly against the patient's chest wall over the intercostal spaces. Try not to listen directly over bone. Never listen through clothing, which impedes or alters sound transmission. Avoid other extraneous sounds, such as those caused by the stethoscope rubbing against bed rails or other objects. Keep the stethoscope still and in place.

AUSCULTATION TIP *If the patient has a hairy chest or back, lightly dampen the chest and hold the stethoscope firmly against the skin to minimize the crackling noises produced by the hair.*

AUSCULTATORY SEQUENCE

The auscultatory sequence for the posterior chest wall surface includes 10 different sites. (See *Posterior auscultatory sequence,* page 122.) The first site is above the left scapula over the lung apex. From there, the auscultatory sequence follows a pattern that progresses downward from the lung apices to the bases. This pattern covers the entire posterior chest wall surface and includes a comparison of sounds heard over the same auscultatory site over both the right and left lungs. Sites 5, 7, and 10 are located over the lateral chest wall surfaces.

The anterior chest wall auscultatory sequence includes nine sites and follows the same pattern as the posterior chest wall sequence. The pattern also includes sites over the lateral chest wall surfaces. (See *Anterior auscultatory sequence,* page 122.)

COMPARING BREATH SOUNDS

If the patient is alert and healthy, begin auscultation with the patient sitting upright and leaning slightly forward. Position yourself behind the patient. Ask the patient to breathe through an open mouth, slightly deeper than usual, through several respiratory cycles. (*Note:* If the patient is extremely dyspneic, don't ask him to take deeper breaths. In this case, begin auscultating for breath sounds at the bilateral lung bases.) Place the diaphragm of the stethoscope over the left lung apex, and listen for at least one complete respiratory cycle. Then move the diaphragm to the same site over the right lung. Compare the breath sounds heard over these

Posterior auscultatory sequence

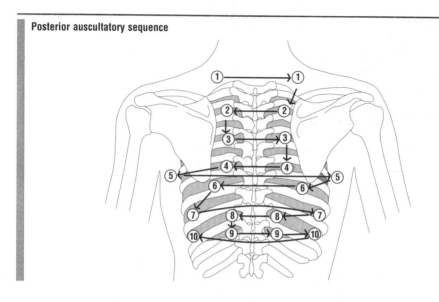

Anterior auscultatory sequence

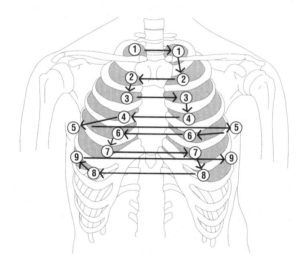

same locations. Continue in this manner, making contralateral comparisons at each auscultatory site.

After you have auscultated the entire posterior chest wall and parts of the lateral chest walls, move to the front of the patient. Have the patient place his arms at his sides and breathe through an open mouth, slightly deeper than usual, as you listen to breath sounds over the anterior chest wall. You can obtain additional information about tracheal sounds by auscultating over the sternum, larynx, and mouth.

AUSCULTATION TIP *You may roll patients who are comatose, critically ill, or bedridden from one side to another to auscultate dependent lung regions. Listen initially over dependent lung regions because gravity-dependent secretions or fluids may produce abnormal sounds that sometimes disappear when the patient is turned, breathes deeply, or coughs.*

AUSCULTATION TIP *To gain the trust and cooperation of a child, allow him to listen to your lungs or to a stuffed animal or doll.*

◀◀◀ POSTTEST

1. What is the primary function of the respiratory system?
2. Name the anatomic structures of the upper airway.
3. Name the three sections of the pharynx.
4. Name the anatomic structures of the lower airway.
5. Where is the trachea located?
6. Describe the anatomy of the lower airway from the mainstem bronchi to the alveoli.
7. What is anatomic dead space?
8. What are the functions of the mucus layer and the cilia in the airway?
9. How are the lungs separated into lobes?
10. Where are the apex and base of the lung located?
11. What are the three types of cells found in the alveoli?
12. What is surfactant and where is it found?
13. What roles do the parasympathetic and sympathetic nervous systems have in the lower airways?
14. Describe the pulmonary circulation in the lungs.
15. Describe the bronchial circulation in the lungs.
16. What are the pleurae?
17. What is the function of the thorax during breathing?
18. What is the primary muscle of respiration and how does it contribute to breathing?
19. How do the external intercostal muscles contribute to breathing?
20. Name the accessory muscles of respiration, and describe their role in breathing.
21. Describe normal breathing.
22. How are air and blood distributed in the lungs?
23. Describe how postural changes affect ventilation and blood flow.

24. How do the tracheobronchial tree's dimensions and shape influence airflow patterns?
25. Name the anatomic structures and landmark lines on the anterior chest wall surface.
26. Name the anatomic structures and landmark lines on the lateral chest wall surface.
27. Name the anatomic structures and landmark lines on the posterior chest wall surface.
28. What are the functions of the diaphragm and bell of the stethoscope?
29. How is the stethoscope used to auscultate for breath sounds?
30. Describe the sequence used during auscultation to compare breath sounds.

Introduction to breath sounds

13

PRETEST ▶▶▶

1. How are breath sounds produced?
2. What are airway dynamics?
3. Name three different airflow patterns.
4. How is the frequency of vibrations measured?
5. Explain the concept of sound damping.
6. Explain the concept of impedance matching.
7. What are egophony, bronchophony, and whispered pectoriloquy?
8. What are adventitious sounds?
9. Name the four characteristics of breath sounds that should be documented in a patient's record.

Breath sound production

Three types of sounds are heard during auscultation — normal breath sounds, abnormal breath sounds (including voice sounds), and adventitious sounds. Breath sounds are produced by airflow patterns, by associated pressure changes within the airways, and by solid tissue vibrations. Both normal and abnormal breath sounds have certain recognizable characteristics. Their amplitude and intensity are believed to be affected by airflow patterns, regional lung volume or distribution of air, body position, and the sound production site.

The sounds are normally diminished and filtered when they are transmitted through air-filled alveoli, fluid accumulations in the pleura, and solid structures, such as bone.

BREATH SOUND ORIGINATION

The sites where normal breath sounds originate aren't clearly defined. For example, sounds heard over the trachea and large airways are characterized as loud and tubular (hollow) with a long expiratory phase. These sounds are thought to reflect the turbulent airflow patterns in the first divisions of the large airways.

Sounds heard over other chest wall areas are softer and have a shorter expiratory phase. These sounds may also be produced by turbulent airflow in the first few divisions of the large airways; however, the fact that these sounds vary with changing airflow rates and distribution of ventilation in the lungs suggests that they are produced in the peripheral airways. Also, inspiratory sounds may be produced in the peripheral airways, and expiratory sounds may be produced in the larger airways. Breath sounds aren't produced in the terminal bronchioles because of minimal airflow rates.

DISEASE

Disease processes that alter the airway or airflow dynamics produce abnormal breath sounds or adventitious sounds. These sounds are generated by the vibration of solid structures, by airflow through narrowed airways, or by abrupt changes in airway pressure.

AIRWAY DYNAMICS

The stability of an airway depends on the interaction of the elastic properties of the lung tissue, the intrapulmonary airways, and the pressures — both internal and external — exerted on the airways. During the respiratory cycle, a pressure gradient exists across the airway wall and between the trachea and alveoli. As the chest expands during inspiration, intrapleural pressure falls. At the same time, the lung's elastic recoil increases, exerting traction on airway walls. This traction increases the airway's diameter and decreases the resistance to airflow. The pressure gradient between the alveolar and atmospheric pressures increases, driving air through the airways toward the alveoli.

AGE ISSUE *In older persons, pulmonary function decreases because of respiratory muscle degeneration or atrophy. Ventilatory capacity diminishes because the lungs' diffusing capacity declines, decreased inspiratory and expiratory muscle strength diminishes vital capacity, and lung tissue degeneration causes a decrease in the lungs' elastic recoil capacity, which results in an elevated residual volume. For these reasons, aging can cause emphysema.*

Intrapleural pressure and alveolar elastic recoil pressures drive expiratory airflow. At the beginning of expiration, intrapleural pressure and the alveolar recoil force cause the alveolar pressures to rise and push the air through the airways and out through the mouth. Air pressures fall as the air flows through the airways from the alveoli to the mouth. A point is reached, particularly during forceful expiration, when intrapleural and intrapulmonary pressures are equal; this normally occurs in the segmental bronchi and is called the *equal pressure point*. Downstream from this point, pressures outside of the airways can compress and further narrow the airways, causing increased resistance to airflow.

AIRFLOW PATTERNS
Variations in airflow patterns are affected by the intricate network of branching airways, the various airway diameters, and the potential for irregular airway wall surfaces within the tracheobronchial tree.

Turbulent airflow
During rapid, or *turbulent*, airflow, air molecules move randomly, colliding against airway walls and each other. They may move across or against the general direction of airflow. The colliding air molecules produce rapid pressure changes within the airway, and these rapid pressure changes produce sounds. Turbulent airflow occurs in the trachea, mainstem bronchi, and other larger airways.

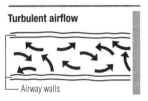

Turbulent airflow

Airway walls

Vortices
As airflow is forced to change directions abruptly in the branching airways, the airstream separates into layers and moves at different velocities. The shearing force of high-velocity airstreams, along with slower airstreams, precipitates opposing circular airflows, or *vortices*. This airflow pattern generates sounds as the flow of air carries the vortices downstream.

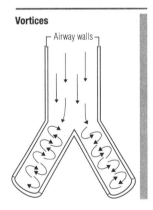

Vortices

Airway walls

Laminar airflow
In the terminal or small airways and respiratory bronchioles, airflow is slow and laminar, or *nonturbulent.* No abrupt changes in pressure or airway wall movements occur to generate sound. Consequently, air movement in these areas produces no sound.

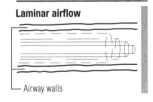

Laminar airflow

Airway walls

MECHANICAL VIBRATIONS

Breath sounds are also produced by mechanical vibrations of solid tissue. These vibrations travel through the different media of air, fluid, or tissue at varying speeds. Their frequency, intensity, and duration can be measured.

Frequency

Frequency, measured in hertz (Hz), refers to the number of vibrations occurring per unit of time. Different frequencies produce the different sounds heard during auscultation. In the clinical setting, the term *pitch* is used to describe sound frequency. High-pitched sounds have higher frequencies; low-pitched sounds, lower frequencies.

Intensity

Intensity, the loudness or softness of the vibrations producing breath sounds, can be measured electronically by recording amplitude. Intensity is affected by the type of vibrating structure producing the sound, the distance the sound travels, and the transmission pathway.

Duration

The duration of the vibrations that produce breath sounds can be measured in milliseconds (msec). However, the human ear perceives sounds during auscultation as long or short and as continuous or discontinuous.

SOUND DAMPING

Body structures or cavities that produce resonance selectively transmit or amplify breath sounds that are similar in frequency and absorb, or damp, those with other frequencies. Breath sounds arising from the same location in the lung have a higher pitch when heard at the mouth or over the trachea than when heard over the chest wall. This change in pitch occurs because high-pitched breath sounds are absorbed as they're transmitted through the lungs and thorax.

Voice sounds are also affected by damping. The resonant qualities of the mouth, nasopharynx, paranasal sinuses, and chest cavity contribute to the pitch and intensity of voice sounds.

IMPEDANCE MATCHING

Breath and voice sounds are normally damped when they're transmitted through air, fluid, and tissue. However, when these sounds are transmitted sequentially through media with different acoustic properties, the sounds' vibrations are filtered and reflected at the interface of the different media. The amount of filtering that takes place depends on the media's impedance (resistance to sound transmission).

Impedance match
When two media are matched according to acoustic properties, sound is transmitted effectively. For example, consolidated areas enhance the transmission of breath sounds to the chest wall. This happens because the consolidating substance, an inflammatory exudate, collects in alveoli, replacing air with dense lung tissue. Consequently, the fluid-filled, airless lung tissue and the chest wall are acoustically well matched, and breath sounds are transmitted more easily.

Impedance mismatch
The transmission of breath and voice sounds through either tissue or fluid between inflated lung segments and the chest wall results in an impedance mismatch. The vibrations of the sounds are significantly filtered and reflected back to their source by the pleurae. Diseases that increase the impedance mismatch, such as a large pleural effusion or empyema, diminish breath sound intensity. Obesity is also associated with diminished breath sounds because of the increased distance between the stethoscope and the lungs.

AUSCULTATION TIP *For better results when auscultating the lungs of a patient who's obese, ask him to take deep breaths through his open mouth while sitting upright or standing.*

Terminology

CLASSIFYING BREATH SOUNDS
Normal breath sounds are classified as tracheal and mainstem bronchi sounds and normal breath sounds heard over other chest wall areas. The tracheal and mainstem bronchi sounds, produced by turbulent airflow, are loud and can be heard throughout inspiration and expiration over the trachea and mainstem bronchi. (♦2-1) Normal breath sounds heard over other chest wall areas are faint and can be heard throughout inspiration and at the beginning of expiration. (♦2-2) (See Chapter 14 for further discussion of normal breath sound terminology.)

Bronchovesicular breath sounds represent a third classification of normal sounds. These sounds are heard over areas between the mainstem bronchi and the smaller airways; their pitch and duration are midway between those of tracheal and mainstem bronchi breath sounds and normal breath sounds heard over other chest wall areas. (◆2-3) Bronchovesicular sounds are heard during inspiration and expiration for equal amounts of time.

CLASSIFYING VOICE SOUNDS
Voice sounds are vibrations produced by speech that are transmitted to the chest wall through the tracheobronchial tree. Voice sounds are classified as bronchophony, egophony, and whispered pectoriloquy.

Voice sounds auscultated over the chest wall are called *bronchophony.* In a healthy individual, bronchophony is similar to voice sounds heard through the neck. (◆2-4) Voice sounds transmitted through the chest wall that have selectively amplified higher frequencies are called *egophony.* (◆2-5) High-pitched whispered sounds transmitted through airless, consolidated lung tissue are called *whispered pectoriloquy.* (◆2-6)

CLASSIFYING ADVENTITIOUS SOUNDS
Adventitious sounds are added sounds that are heard with normal and abnormal breath sounds. They're classified as either crackles or wheezes and then further classified by duration and pitch. (See Chapter 18 for a further discussion of adventitious breath sound terminology.)

Documenting auscultation findings

Careful documentation of auscultation findings is necessary to determine if sounds heard have changed over time. This requires the same systematic approach used during auscultation. First, determine the location, intensity, duration, and pitch of each sound during auscultation, and then document these characteristics in the patient's record.

LOCATION
Document sound *location* by describing anatomic thoracic landmarks and indicating the anterior, posterior, or lateral chest wall surface where the sound is heard.

AGE ISSUE *Because the chest wall of the neonate is very thin, sound is transmitted throughout all lung fields, making it difficult to pinpoint its origin.*

You can also document location by using the lung bases or apices as reference points. Also document whether the sounds are heard bilaterally (over both lungs) or unilaterally (over one lung).

INTENSITY

Sound *intensity* is described subjectively, using such terms as *loud, soft, absent, diminished,* or *distant.*

DURATION

Sound *duration* refers to the sound's timing within the respiratory cycle—that is, whether it's heard during inspiration, expiration, or both. Timing can be described as early, late, or throughout.

PITCH

Sound *pitch* is usually described as high or low. Documenting pitch is more important when wheezes or crackles have been noted during auscultation because pitch helps differentiate the underlying disease process.

PUTTING THEM ALL TOGETHER

A typical documentation example might read: "Late inspiratory fine crackles were heard over the right base posteriorly along the midscapular line, and coarse crackles were heard throughout inspiration and expiration over the left apex anteriorly near the midclavicular line." Such a description provides the reader with valuable information about any adventitious sounds heard during auscultation.

◀◀◀ POSTTEST

1. How are breath sounds produced?
2. Briefly describe the differences between breath sounds heard over the trachea and mainstem bronchi and those heard over other chest wall areas.
3. What role do pressure gradients have in breathing?
4. Describe how intrapleural pressure and alveolar elastic recoil pressures serve as the driving pressure for expiratory flow.
5. What influences airflow patterns?

6. Describe turbulent airflow.
7. Describe vortices.
8. Describe laminar airflow.
9. What is meant by frequency, intensity, and duration of breath sounds?
10. Explain the concept of sound damping.
11. Explain the concept of impedance matching.
12. What happens to impedance matching when the lungs become air- or fluid-filled?
13. Compare and contrast tracheal and mainstem bronchi breath sounds with normal breath sounds heard over other chest wall areas.
14. What are bronchovesicular breath sounds?
15. Define the terms *bronchophony, egophony,* and *whispered pectoriloquy.*
16. What are adventitious sounds?
17. Name the four characteristics used to describe breath sounds.
18. Which terms are used to describe sound intensity?
19. Which terms are used to describe sound duration?
20. Which terms are used to describe sound pitch?

FIVE

Normal breath sounds

Breath sounds heard in healthy individuals

14

PRETEST ▶▶▶

1. What are normal breath sounds?
2. How do normal breath sounds originate?
3. What affects normal breath sound characteristics?
4. What is the significance of breath sounds heard at the mouth?

Anatomy and physiology

Normal breath sounds are broadly defined as breathing heard through the chest wall of a healthy person. They include tracheal, bronchial, vesical, and bronchovesicular sounds, as well as sounds heard at the mouth. Three variables affect the normal sounds heard during auscultation: the distance between the source of the sound and the chest wall, the path of sound transmission, and the sound's location. Normal breath sounds are produced by airflow patterns, by associated pressure changes within the airways, and by solid tissue vibrations.

Normal breath sounds heard over large airways are produced by turbulent airflow in those airways. They're loudest when auscultated over the trachea on the anterior chest wall surface next to the sternum. As air travels through the airways beyond the segmental bronchi, these normally high-pitched sounds become diminished because they're filtered by the chest wall, pleurae, and air-filled lung tissue. The normal breath sounds heard over most of the chest wall surface are soft and low pitched; they're softer and shorter during expiration than during inspiration.

Auscultatory sites for normal breath sounds heard over other chest wall areas

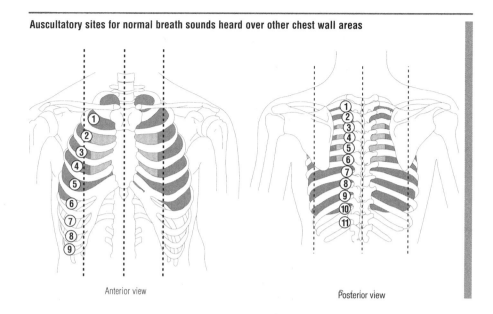

Anterior view

Posterior view

TRACHEAL AND BRONCHIAL BREATH SOUNDS

Normal breath sounds heard over the trachea (tracheal breath sounds) and mainstem bronchi (bronchial breath sounds) are produced by turbulent airflow patterns. Tracheal breath sounds are described as harsh and high pitched and can be heard over the trachea. Bronchial breath sounds are described as loud and high pitched and can be heard next to the trachea. Inspired and expired air travels easily through these large airways.

When tracheal breath sounds are being auscultated, a brief pause may be heard at the end of inspiration, and expiration is longer than inspiration. The inspiratory-expiratory (I:E) ratio is 1:2 to 1:3. The sound frequency distribution of these breath sounds is 200 to 2,000 Hz. **(◆2-7)**

Tracheal and mainstem bronchi (bronchial) breath sounds are heard over the chest wall on either side of the sternum from the second intercostal space to the fourth intercostal space anteriorly and along the vertebral column from the third intercostal space to the sixth intercostal space posteriorly. These sounds have a variable pitch and a loud intensity heard best using the diaphragm of the stethoscope; they're heard throughout inspiration and expiration. **(◆2-8)** (See *Auscultatory sites for tracheal and mainstream bronchi breath sounds,* page 136.)

Airway area producing tracheal and mainstream breath sounds

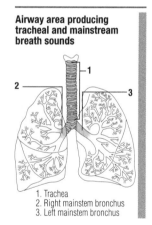

1. Trachea
2. Right mainstem bronchus
3. Left mainstem bronchus

Auscultatory sites for tracheal and mainstream bronchi breath sounds

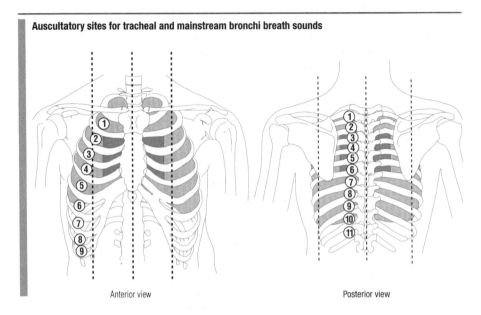

Anterior view   Posterior view

Airway areas producing vesicular breath sounds

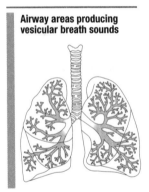

VESICULAR BREATH SOUNDS

Vesicular breath sounds are transmitted through lung tissue and the chest wall and are produced by changes in airflow patterns. These normal sounds are quieter than tracheal and bronchial sounds. Inspiration is heard clearly, and expiration immediately follows inspiration but quickly fades as airflow rates rapidly decline and turbulent airflow is directed toward the central airways. The I:E ratio is 3:1 to 4:1. The sound frequency distribution of these normal breath sounds is 200 to 600 Hz. (◆2-9)

BRONCHOVESICULAR BREATH SOUNDS

Bronchiovesicular breath sounds are heard anteriorly and posteriorly over the central large airways. Their pitch and duration are midway between those of the vesicular breath sounds and the bronchial sounds; inspiratory and expiratory phases are equal, with an I:E ratio of 1:1.

❀ **AGE ISSUE** *Bronchovesicular breath sounds may be heard in the outer fields in children as old as age 6 because infants and young children have thin chest walls and muscles that don't diminish breath sounds to the same extent as the thicker chest walls of adults.*

Variations in intensity

The intensity of bronchovesicular breath sounds changes during deep inspiration and after maximal expiration. Intensity may also change depending on the lung area being auscultated. For example, if the patient is sitting up during auscultation of the anterior chest wall surface, breath sounds are louder over the apices during early inspiration and become progressively softer as inspiration continues. (♦2-10) During auscultation of the posterior lung bases, breath sounds are initially softer and become progressively louder during maximal inspiration.

Variations in intensity are usually related to airflow patterns and to the distribution of air throughout the lungs. During inspiration with the patient in the upright position, air flows initially into the lung apices, then into the lung bases (the dependent areas) after the small airways reopen; therefore, intensity will probably be greater in the apices than in the bases. During normal breathing, this change in intensity may not always be audible.

Variations in intensity that are unrelated to airflow rates or regional ventilation may be heard when comparing symmetrical sites on the chest wall. For example, breath sounds heard over the left lower lobe may vary in intensity with the cardiac cycle. Intensity may increase during systole because ventricular contraction allows the surrounding lung tissue to expand more fully, increasing turbulent airflow to that region and intensifying the inspiratory sounds. Conversely, intensity may decrease during diastole because ventricular expansion of the heart compresses adjacent lung tissue, reducing airflow to that region.

Bronchovesicular breath sounds have a low pitch that can be heard using either the diaphragm or bell of the stethoscope. They're softer and less harsh than bronchial breath sounds and have a tubular quality. (♦2-11)

NORMAL BREATH SOUNDS HEARD AT THE MOUTH

Normal breath sounds heard at the mouth are thought to be produced by turbulent airflow occurring below the glottis and before the terminal airways. In healthy individuals, breath sounds at the mouth provide baseline data that can be useful later when evaluating noisy or paradoxically quiet breath sounds.

Normal breath sounds heard at the mouth are heard at the lips. Their intensity is as loud and harsh as that of tracheal and mainstem bronchi breath sounds, and their duration is continuous (throughout the respiratory cycle). They have a moderately high pitch.

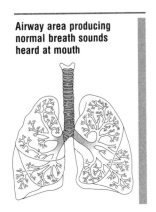

Airway area producing normal breath sounds heard at mouth

◀◀◀ **POSTTEST**

1. Define normal breath sounds.
2. What variables affect normal breath sounds heard during auscultation?
3. How are normal breath sounds produced?
4. Describe tracheal and mainstem bronchi breath sounds.
5. Describe breath sounds heard over other chest wall areas.
6. How do ventilation and airflow rate changes affect breath sound intensity?
7. How do cardiac systole and diastole affect normal breath sound intensity?
8. How are breath sounds heard at the mouth produced?

SIX

Abnormal breath sounds

Bronchial breath sounds
15

PRETEST ▶▶▶

1. What happens to normal breath sounds when lung density increases?
2. What are bronchial breath sounds?
3. How does consolidation affect breath sounds?
4. How does atelectasis affect breath sounds?
5. How does fibrosis affect breath sounds?

Sound production

When lung tissue between the central airways and the chest wall becomes airless because of conditions that increase lung density, transmission of breath sounds from large airways is enhanced. This happens because little high-frequency sound is lost through attenuation or filtration. As lung tissue density increases the impedance between the fluid-filled lung tissue and the pleurae and chest wall, these three media become well matched, which decreases the normal filtering of high-frequency sounds. Consequently, breath sounds are transmitted more readily to the chest wall surface and are louder and more tubular than normal breath sounds heard over the same chest wall area. Expiration is significantly louder and longer than normal. The inspiratory-expiratory (I:E) ratio changes from the normal 3:1 or 4:1 to 1:1 or 1:2. These sounds, called *bronchial breath sounds,* are considered abnormal when found anywhere except anteriorly over the large airways.

RELATED CONDITIONS
Conditions associated with bronchial breath sounds include consolidation, atelectasis, and fibrosis, all of which increase lung tissue density because of fluid accumulation, lung col-

lapse, or fibrotic scarring. Bronchial breath sounds are
heard over the affected lung area.

Consolidation

The most common cause of lung tissue consolidation (solidification) is pneumonia, a lung inflammation that can be
caused by bacteria, viruses, or chemical insults (such as
with aspiration). In this condition, fluid, leukocytes, and
erythrocytes accumulate in spaces that are normally air
filled, producing a consolidated area. Clinical findings vary,
depending on the location of the consolidated area and the
causative agent. When classic consolidation is present, decreased chest wall movement and a dull percussion note are
apparent over the affected area. Bronchial breath sounds are
heard over a dense, airless *upper* lobe, even without a
patent bronchus (◆**2-12**), because the upper lobe surfaces
are in direct contact with the trachea and loud tracheal
breath sounds are transmitted directly to the dense, airless
upper lobe tissues. In contrast, bronchial breath sounds are
heard over a dense, airless lower lobe only when the
bronchi are patent, because sound isn't transmitted directly
to the airless lower lobe tissues.

Sound characteristics

In a patient with lobar pneumonia with right posterior midlung consolidation, bronchial breath sounds are heard over
the right posterior midlung field, located over the seventh
and eighth intercostal spaces along the vertebral column.
(See *Bronchial breath sounds heard over consolidated right
posterior midlung,* page 142.) These sounds are high-
pitched and have the typical hollow, or tubular, quality of
normal central airway breath sounds. They remain audible
during both expiration and inspiration, but the expiratory
sounds are longer and louder when the patient is sitting up;
the I:E ratio is 1:2. These sounds may be auscultated using
either the bell or diaphragm of the stethoscope. (◆**2-13**)

Atelectasis

Atelectasis, incomplete expansion of a lung area, is typically diagnosed in postoperative or immobile patients and in
some patients with bronchiectasis or pneumonia. It's
thought to result from prolonged shallow breathing (hypoventilation) or uncleared secretions that occlude the airway. Because no air enters the distal airways, the segmental
or lobar bronchi collapse. If a large airway is occluded,

Bronchial breath sounds heard over consolidated right posterior midlung

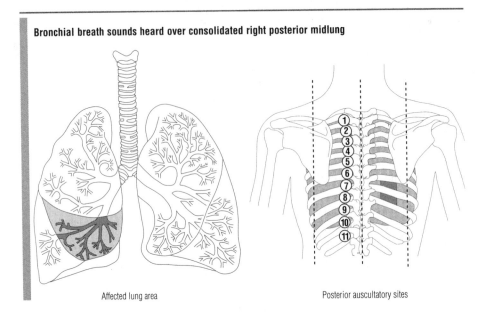

Affected lung area   Posterior auscultatory sites

clinical findings include decreased chest wall movement, a dull percussion note, regional changes in lung volume, and bronchial breath sounds over a dense, airless upper airway. (◆2-14) However, decreased chest wall movement and lung volume changes may be difficult to detect on clinical examination.

ALERT *If you hear bronchial breath sounds in upper lobes and absent breath sounds in lower lobes, assume that something is occluding the large airway. Provide measures to maintain oxygenation.*

Sound characteristics
Bronchial breath sounds are heard over an atelectatic area when the bronchus is patent. They aren't heard over an atelectatic lower lobe if the bronchus is obstructed.

In an unresponsive patient with a head injury, bronchial breath sounds are heard over the right anterior midlung field, located between the third and fifth intercostal spaces from the midsternal line to just right of the midclavicular line. These sounds are high pitched and have the typical hollow, or tubular, quality of normal central airway breath sounds. They're audible throughout inspiration and expiration. When the patient is in the supine position, the inspiratory and expiratory sounds are equal in duration and intensity; the I:E ratio is 1:1. These sounds can be heard equally

Bronchial breath sounds heard over atelectatic right anterior midlung

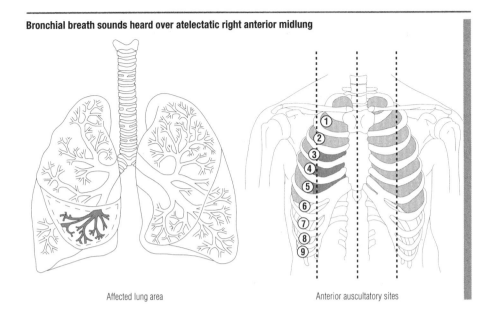

Affected lung area　　　　　　　　　　Anterior auscultatory sites

well using either the bell or diaphragm of the stethoscope. (◆2-15)

Fibrosis

Severe fibrosis (abnormal formation of fibrous connective tissue) may produce bronchial breath sounds that are usually heard over the lower lung regions. Interstitial pulmonary fibrosis is a pathologic change caused by many chronic inflammatory diseases that produce diffuse lung injury. Some possible causes of pulmonary fibrosis include excessive smoke inhalation and excessive exposure to asbestos. In most cases, however, the cause is unknown. The breath sounds heard over fibrotic areas are similar to those heard over atelectatic areas. (◆2-16)

Sound characteristics

In some patients with asbestosis, bronchial breath sounds are heard anteriorly and posteriorly over both lung bases.

　　Anteriorly, this area extends from the midsternal line to the right and left of the anterior axillary lines over the fifth and sixth intercostal spaces; posteriorly, it extends from the vertebral line to the right and left of the posterior axillary lines over the seventh, eighth, ninth, and tenth intercostal spaces. (See *Auscultatory sites for asbestosis in lung bases,* page 144.)

Lung area affected by asbestosis

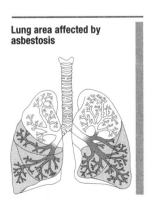

Auscultatory sites for asbestosis in lung bases

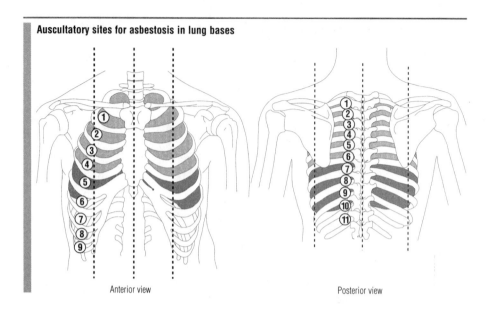

Anterior view   Posterior view

Because the bronchi are patent, bronchial breath sounds heard over fibrotic areas have the typical hollow, or tubular, quality of normal central airway breath sounds. They're audible throughout inspiration and expiration. When the patient is in the upright position, the breath sounds become progressively louder during inspiration and become both louder and longer during expiration; the I:E ratio is 1:1 to 1:2. These bronchial breath sounds are high pitched and are heard equally well with either the diaphragm or bell of the stethoscope. (◆2-17)

◀◀◀ POSTTEST

1. What happens when lung tissue becomes airless because of conditions that increase lung density?
2. Why are breath sounds enhanced over an area of increased lung density?
3. Name three conditions associated with bronchial breath sounds.
4. Which clinical findings are typically associated with consolidation?
5. What role does a patent bronchus have in the transmission of breath sounds to the apex and base of the lung?
6. Describe the characteristics of bronchial breath sounds heard over a consolidated area.

7. What causes atelectasis?
8. Describe the characteristics of bronchial breath sounds heard over an atelectatic area.
9. Describe the characteristics of bronchial breath sounds heard over a fibrotic area.

Abnormal voice sounds
16

PRETEST ▶▶▶

1. How are normal voice sounds produced?
2. Name three voice sounds.
3. How do voice sounds change over areas of dense, airless lung tissue?
4. Which specific words or letters is the patient asked to repeat to detect changes in bronchophony and egophony?
5. Which auscultatory findings are associated with consolidation?

Sound production

Voice sounds are produced by vibrations of the vocal cords as air from the lungs passes over them. The resonance of the mouth, nasopharynx, and paranasal sinuses helps to amplify these sounds. When voice sounds pass through normally inflated, air-filled lungs, vowel tones, which contain many high-frequency sounds, are diminished and filtered. Like other high-frequency sounds, voice sounds are also selectively reflected back toward the lung tissue at the pleurae. Consequently, transmitted voice sounds are normally heard as a low-pitched, unintelligible mumble over healthy lung and pleural surfaces because most of the vowels are filtered out; conversely, they're heard distinctly over consolidated or atelectatic lung tissue areas because less filtering takes place, enhancing transmission.

146

Auscultatory sites for bronchophony heard over consolidated left upper lobe

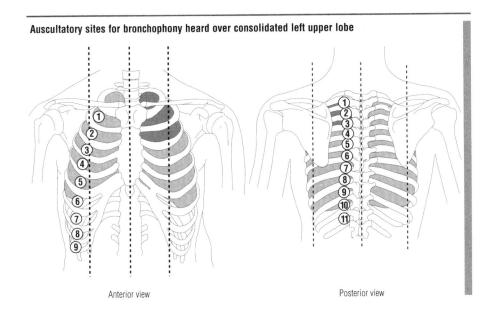

Anterior view
Posterior view

THREE TYPES

The three types of voice sounds are bronchophony, whispered pectoriloquy, and egophony.

Bronchophony

Bronchophony is the clear, distinct, intelligible voice sound heard over dense, airless lung tissue. (◆2-18) Dense, airless lung tissue transmits high-frequency vowel sounds more easily because of impedance matching. Also, consolidation increases vocal resonance, which allows the clear transmission of voice sounds to the chest wall. Bronchophony is heard over dense, airless upper lobes because the upper lobe surfaces are in direct contact with the trachea, leading to direct transmission of tracheal breath sounds. In contrast, bronchophony is heard over dense, airless lower lobes only when the bronchi are patent because a direct path of sound transmission to the lower lobes doesn't exist.

Sound characteristics

Bronchophony can be heard anywhere over the anterior, lateral, and posterior chest wall surfaces. It's commonly heard over dense, airless lung tissue in the upper lobes, such as a consolidated area. In a patient with left upper lobe pneumonia, intelligible voice sounds are heard over both the anterior and the posterior chest wall surfaces. The anterior area is

Consolidated lung area producing bronchophony in left upper lobe

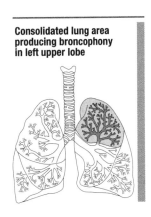

Auscultatory sites for whispered pectoriloquy heard over atelectatic left lower lobe

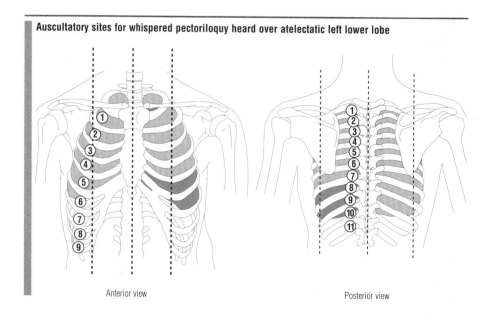

Anterior view

Posterior view

located between the space above the clavicle and the second intercostal space from the midsternal line to the left of the midclavicular line. The posterior area is located between the first and third intercostal spaces from the vertebral line toward the 1eft midscapular line.

The patient is asked to repeat the words "ninety-nine" several times. Over healthy lung tissue, the words are unintelligible **(◆2-19)**; over a consolidated area, however, the high-frequency sounds are easily understood as words. **(◆2-20)**

Whispered pectoriloquy

Whispered pectoriloquy is the clear, distinct, intelligible whispered voice sound heard over airless, consolidated, or atelectatic lung tissue. In a healthy person, normal lung tissue filters the high frequencies of whispered vowel sounds, making them unintelligible during auscultation. In a patient with consolidation or atelectasis, these same whispered vowel sounds are transmitted to the chest wall surface without much filtering of high frequencies and can be heard clearly with the stethoscope. **(◆2-21)**

Sound characteristics

Whispered pectoriloquy can be heard anywhere over the anterior or posterior chest wall surfaces over dense, airless

Atelectatic lung area producing whispered pectoriloquy in lower left lobe

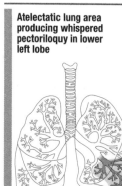

lung tissue, typically over an area of consolidation or atelectasis. In a postoperative patient with left lower lobe atelectasis and a patent bronchus, intelligible voice sounds are heard over the anterior and posterior chest wall surfaces. The anterior area is located over the fifth and sixth intercostal spaces from the midsternal line to the midaxillary line. The posterior area is located over the eighth, ninth, and tenth intercostal spaces from the vertebral line to the midaxillary line.

The patient is asked to whisper words, such as "one-two-three" several times. Over healthy lung tissue, the words are unintelligible **(◆2-22)**; over the atelectatic area, the high-frequency vowel sounds are easily understood as words. **(◆2-23)**

Egophony

Egophony, a voice sound that has a nasal or bleating quality when heard over the chest wall **(◆2-24)**, is produced over consolidated and atelectatic areas because impedance matching enhances breath sound transmission in such areas. It's also detectable over the upper border of large pleural effusions.

Sound characteristics

Egophony can be heard anywhere over the anterior, posterior, or lateral chest wall surfaces over a consolidated or atelectic area. In a patient with right lower lobe pneumonia, egophony is heard over the lateral chest wall surface. This area is located over the fifth and sixth intercostal spaces between the anterior axillary line and the midaxillary line.

The patient is asked to say the letter "E" several times. Over healthy lung tissue, the letter sounds as it normally does **(◆2-25)**; over the consolidated area, it sounds like "A-aay" and has a high-pitched, nasal quality. **(◆2-26)**

CONSTELLATION OF FINDINGS

A constellation of auscultatory findings—bronchial breath sounds, bronchophony, whispered pectoriloquy, and egophony—is typical in patients with consolidation. Bronchophony may be easier to hear when bronchial breath sounds are pronounced and consolidation is present in only one lobe or lung. Whispered pectoriloquy is typically discernible over atelectatic lung segments or areas of patchy consolidation where bronchial sounds or bronchophony isn't completely audible.

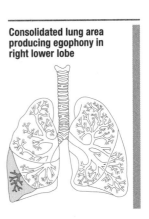

Consolidated lung area producing egophony in right lower lobe

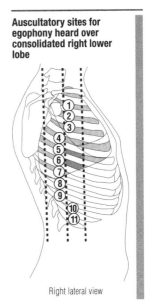

Auscultatory sites for egophony heard over consolidated right lower lobe

Right lateral view

◀◀◀ POSTTEST

1. How are voice sounds produced?
2. Describe voice sounds heard over normal lung and pleural surfaces.
3. What happens to voice sounds heard over a consolidated area or an atelectatic area?
4. How does bronchophony change over dense, airless lung tissue?
5. What role does a patent bronchus have in transmitting bronchophony to the lung's apex and base?
6. Describe the characteristics of bronchophony.
7. Which words is the patient asked to repeat to detect changes in bronchophony?
8. How does whispered pectoriloquy change over dense, airless lung tissue?
9. Describe the characteristics of whispered pectoriloquy.
10. Which words is the patient asked to repeat to detect changes in whispered pectoriloquy?
11. How is egophony produced over dense, airless lung tissue?
12. Describe the characteristics of egophony.
13. Which letter is the patient asked to repeat to detect changes in egophony?
14. What constellation of auscultatory findings is typical in patients with consolidation?

Absent and diminished breath sounds

17

PRETEST ▶▶▶

1. Which conditions are associated with diminished or absent breath sounds?
2. How does each of these conditions cause diminished or absent breath sounds?
3. Explain the concepts of impedance mismatch and filtering.
4. Does lung hyperinflation cause diminished or absent breath sounds?
5. How does positive end-expiratory pressure change normal breath sound characteristics?
6. Does obesity change normal breath sound characteristics?

Sound production

Breath sounds are diminished or eliminated by conditions that limit airflow into lung segments. Slow inspiratory airflow rates decrease breath sound amplitude because less air movement occurs, resulting in less turbulent airflow.

Diminished or absent breath sounds can also occur when breath sounds are reflected at the visceral and parietal pleurae because of an impedance mismatch. A mismatch occurs when sounds are transmitted through two types of media with significantly different acoustical properties. For example, when breath sounds, which are normally transmitted through aerated lung tissue, are transmitted through a collection of fluid or air in the pleural space, sound transmission is halted or filtered. The same acoustical mismatch also occurs in patients with increased chest wall thickness. For example, in obese patients, breath sounds are diminished because they're transmitted through a layer of fat.

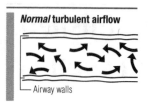

Normal turbulent airflow

Airway walls

Decreased turbulent airflow in shallow breathing

Airway walls

RELATED CONDITIONS

Conditions associated with absent or diminished breath sounds include shallow breathing, diaphragmatic paralysis, severe airway obstruction, pneumothorax, pleural effusion, hyperinflated lungs, and obesity. The use of positive end-expiratory pressure (PEEP) during assisted ventilation also is associated with diminished breath sounds.

Shallow breathing

During normal breathing in the upright position, a certain amount of air flows through the airways during inspiration and expiration; the distribution of ventilation is greater in dependent lung regions because more respiratory movement occurs. During shallow breathing, less respiratory movement occurs; consequently, less air flows through the airways during inspiration and expiration. Because of this reduced airflow, turbulence is decreased and breath sounds are diminished. **(◆2-27)**

Sound characteristics
Diminished breath sounds may be heard over the anterior, posterior, and lateral chest wall surfaces. Postoperative patients and patients with rib fractures commonly breathe shallowly because pain limits their depth of respiration. Patients with decreased levels of consciousness from central nervous system injuries or drug overdoses may also have shallow breathing.

Diaphragmatic paralysis

During inspiration in a healthy person, contraction of the dome-shaped diaphragm expands the lower rib cage, forcing the abdominal contents downward and out and increasing longitudinal lung size. Thoracic expansion lowers intrapulmonary pressure, which allows air to flow into the airways.

When the diaphragm becomes paralyzed, as might happen with injury to the phrenic nerves, it no longer participates in normal breathing. The internal and external intercostal muscles, which also have a role in normal breathing, must take over the work of breathing. With only the chest wall muscles initiating the respiratory cycle, ventilation of the lung bases may be limited, resulting in diminished breath sounds.

Lobar bronchus obstruction

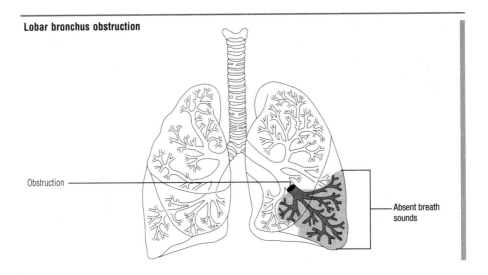

Obstruction

Absent breath sounds

Sound characteristics

Diminished sounds may be heard over the anterior, posterior, and lateral chest wall surfaces. (◆2-28)

 ALERT *In a patient with diaphragmatic paralysis, who has greatly diminished breath sounds, assess for respiratory distress and prepare for intubation and mechanical ventilation.*

Airway obstruction

An obstruction in an airway blocks the flow of air and, therefore, changes the breath sounds heard during auscultation. The location of the obstruction determines where the changes in breath sounds are heard.

Sound characteristics

If a lobar or segmental bronchus becomes obstructed, as in a patient who has aspirated a foreign object or one who has a large mucus plug, airflow ceases distal to the obstruction; therefore, breath sounds are absent over the area distal to the obstruction. If a mainstem bronchus is obstructed, breath sounds are absent throughout the entire affected lung. (See *Mainstem bronchus obstruction,* page 154.)

 ALERT *Aspirated foreign bodies are more likely to lodge in the right mainstem bronchus because it provides a more direct vertical passage than the left. For the same reason, endotracheal tubes are commonly misplaced in the right mainstem bronchus. If breath sounds are absent throughout the left*

Mainstem bronchus obstruction

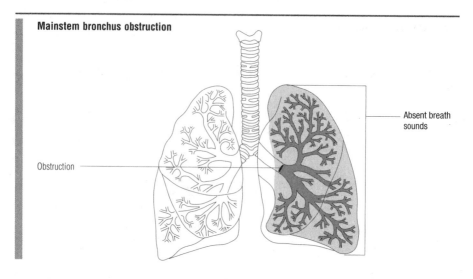

Obstruction

Absent breath sounds

lung field after intubation, the endotracheal tube must be repositioned.

◆ **ALERT** *When a patient has aspirated a foreign object and continues to cough, don't intervene — allow the patient to cough up the object, if possible.*

Pneumothorax

A pneumothorax is the accumulation of air in the normally airless pleural space because of a tear in either the visceral or parietal pleura. Breath sounds heard over a pneumothorax are significantly diminished or absent because of acoustic mismatching of the air-filled lung and the collection of air in the pleural space. **(◆2-29)**

The degree of change in breath sounds depends on the size of the pneumothorax. A small pneumothorax may not alter the breath sound intensity enough to be heard during auscultation; it may, however, be apparent on a chest X-ray. A large pneumothorax significantly diminishes breath sounds or permits no sound transmission, causing absent breath sounds. Typical pneumothorax characteristics include sharp, stabbing chest pain; dyspnea; absent breath sounds; inaudible egophony, bronchophony, and whispered pectoriloquy; and increased resonance during percussion.

◆ **ALERT** *pneumothorax that continues to increase in size may be life-threatening, particularly if the patient is being ventilated mechanically. If the increasing intrapleural air isn't evacuated immediately, intrathoracic pressure may increase, causing decreased cardiac output, hypotension, and eventual death.*

Pneumothorax affecting left lateral lung field

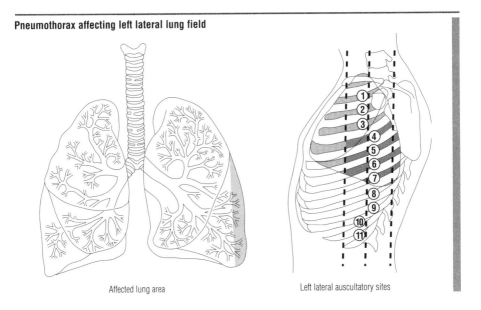

Affected lung area   Left lateral auscultatory sites

Sound characteristics

Diminished or absent breath sounds resulting from a pneumothorax can be heard anywhere over the anterior, posterior, and lateral chest wall surfaces, depending on the location and size of the pneumothorax.

In a patient with a pneumothorax in the left lateral lung field—the area over the fourth, fifth, sixth, and seventh intercostal spaces between the anterior and posterior axillary folds—normal breath sounds are heard on the contralateral side over healthy lung tissue **(◆2-30),** and breath sounds are absent over the pneumothorax. **(◆2-31)** If diminished breath sounds are heard, they have a low pitch that's heard using either the bell or diaphragm of the stethoscope.

Pleural effusion

In pleural effusion, fluid accumulates in the pleural space, impairing the transmission of normal breath sounds. Because of the acoustical mismatch, breath sounds are diminished or absent. **(◆2-32)**

A large pleural effusion compresses adjacent lung tissue, causing atelectasis; the percussion tone is dull. Egophony, bronchophony, and whispered pectoriloquy caused by the atelectasis may be audible at the upper border of the pleural effusion. Occasionally, loud bronchial breath sounds may have sufficient intensity to be transmitted through a small pleural effusion.

Pleural effusion affecting right lower lobe

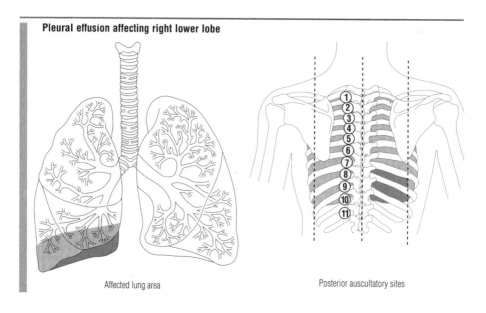

Affected lung area

Posterior auscultatory sites

Sound characteristics

Diminished or absent breath sounds resulting from a pleural effusion may be heard anywhere over the anterior, posterior, or lateral chest wall surfaces, depending on the location of the pleural effusion.

In a patient with a pleural effusion in the right posterior lower lung field, diminished breath sounds are heard over the eighth, ninth, and tenth intercostal spaces from the vertebral line to just right of the midscapular line and normal breath sounds are heard on the contralateral side over healthy lung tissue. (◆2-33) The diminished breath sounds have a low pitch that can be heard using either the bell or diaphragm of the stethoscope. (◆2-34)

Hyperinflated alveoli (hyperinflated lungs)

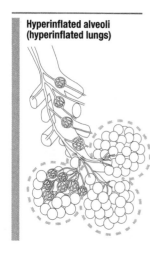

ALERT *If the pleural effusion is large, diminished or absent sounds can be heard over the lower right anterior and lateral chest wall surfaces as well. Monitor the patient for respiratory distress, and prepare for chest tube insertion, if necessary.*

Hyperinflated lungs

Hyperinflated lungs, typical of patients with chronic obstructive pulmonary disease (COPD) or severe asthma, impede breath sound transmission in the following way: Premature dynamic compression of the large central airways and the loss of elastic tension in the lung tissue limit airflow

Auscultatory sites for COPD

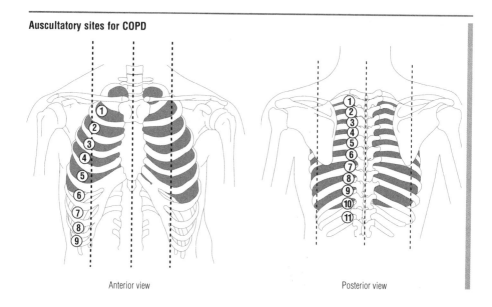

Anterior view   Posterior view

during expiration. Increased intrinsic airway resistance may also be present. Increased amounts of air trapped in the lungs change the acoustical qualities of the sound transmitted, creating a mismatch between the pleurae and chest wall, and the hyperinflated lung tissue. (**◆2-35**) Consequently, breath sounds are diminished.

The intensity of those breath sounds is directly related to the severity of the COPD. For example, patients with marked hyperinflation typically have an increased anterior-to-posterior thoracic diameter and a flattened and immobile diaphragm. Breath sounds aren't as loud as usual because of the increased mismatch produced by trapped air and because less air is being moved during each respiratory cycle. Other signs and symptoms of COPD include a hyperresonant percussion note, dyspnea (a primary sign of severe COPD), prolonged expiratory phase, and inaudible egophony, bronchophony, and whispered pectoriloquy.

ALERT *If a patient with COPD has breath sounds that have progressed from diminished to absent, prepare for intubation and mechanical ventilation.*

Sound characteristics

Diminished or absent breath sounds associated with COPD are heard over the anterior, posterior, and lateral chest wall surfaces. If the breath sounds are audible, they're soft in intensity and they have a low pitch that's heard best using the diaphragm of the stethoscope. These sounds are heard throughout inspiration and expiration. (◆2-36)

Obesity

Breath sounds are normally attenuated and filtered at the pleural surface. In an obese person, a thickened chest wall increases the distance between lung tissue and the chest wall surface, creating an impedance mismatch. This filters breath sounds as they're transmitted from the pleurae to the chest wall surface. (◆2-37)

Sound characteristics

Diminished breath sounds are heard over the anterior, posterior, and lateral chest wall surfaces. The location of fat pads determines where the breath sounds are most difficult to hear. For best results when auscultating an obese patient's lungs, ask him to sit up. Then ask him to take deep breaths through an open mouth while you listen.

Hyperinflated alveoli (PEEP)

Positive end-expiratory pressure

Adding PEEP during mechanical ventilation of an intubated patient can diminish breath sounds in the following way: PEEP increases functional residual capacity (FRC), the amount of air remaining in the airways at the end of normal expiration. Increasing FRC hyperinflates the lungs.

Sound characteristics

Dimished breath sounds are heard because increased amounts of air in the small airways and alveoli change the acoustical qualities of the sound transmitted through the chest wall, creating a mismatch between the pleurae and chest wall, and the hyperinflated lung. (◆2-38)

◀◀◀ POSTTEST

1. What causes diminished or absent breath sounds?
2. Name six conditions associated with diminished or absent breath sounds.
3. Explain why shallow breathing results in diminished breath sounds.
4. Explain the effects of diaphragmatic paralysis on breathing.
5. Describe the changes in airflow distal to an obstruction.

6. Why does a pneumothorax cause diminished or absent breath sounds?
7. Describe the breath sound characteristics associated with a large pneumothorax.
8. What is a pleural effusion?
9. How does a large pleural effusion change breath sounds?
10. Describe the breath sound characteristics associated with a pleural effusion.
11. Why do hyperinflated lungs cause diminished or absent breath sounds?
12. Describe the breath sound characteristics associated with chronic obstructive pulmonary disease.
13. How does obesity change the characteristics of normal breath sounds?
14. Explain the effects of positive end-expiratory pressure on breath sounds.

SEVEN

Other abnormal breath sounds

Classifying adventitious sounds

18

PRETEST ▶▶▶

1. Name two types of adventitious sounds.
2. Describe the difference between fine and coarse crackles.
3. Describe the difference between wheezes and low-pitched wheezes.

Confusion over terminology

Methods of classifying adventitious sounds have changed repeatedly over the years, resulting in continuing confusion in the literature and in the clinical setting. The familiar terms used by Laënnec were replaced by terms that later physicians thought were more aesthetic or clearer. For example, the term *rhonchus* was substituted for *rale* because the latter was associated with the "death rattle" sound. Later, attempts were made to classify adventitious sounds based on their acoustic qualities or their similarities to musical tones.

Current classification system

In 1977, the American Thoracic Society adopted a system of classifying adventitious sounds based on acoustical qualities, timing, and frequency waveforms. This system divides adventitious sounds into two classifications: crackles and wheezes. These two classifications are further subdivided into two categories: discontinuous and continuous.

Discontinuous, explosive sounds that are loud and low pitched are called *coarse crackles*. (♦2-39) Previously, these sounds were called *rales* or *coarse rales*.

Discontinuous, explosive sounds that are shorter in duration, higher in pitch, and less intense than coarse crackles are called *fine crackles*. (◆**2-40**) Previously, these sounds were called *fine rales* or *crepitations*.

Continuous sounds that are high pitched and have a hissing or coughing sound are called *wheezes*. Wheezes commonly have a musical quality. (◆**2-41**) Previously, these sounds were called *sibilant rales* or *sibilant rhonchi*.

Continuous sounds that are low pitched and commonly resemble snoring are called *low-pitched wheezes*. (◆**2-42**) Previously, these sounds were called *sonorous rales* or *sonorous rhonchi*.

It should be noted that the term *rhonchi* is still commonly used in the clinical setting, rather than *low-pitched wheezes*. Rhonchus is used to describe a rough, rumbling low-pitched sound that is heard primarily on expiration, although, at times, rhonchi may also be heard with inspiration. Rhonchi are generally caused by fluid or secretions partially blocking the large airways and typically change in sound, or disappear, with coughing.

◀◀◀ POSTTEST

1. Name the characteristics upon which the American Thoracic Society based its system of classifying adventitious sounds in 1977.
2. What are the two classifications of adventitious sounds?
3. Describe coarse crackles.
4. Describe fine crackles.
5. Describe wheezes.
6. Describe low-pitched wheezes.

Adventitious sounds: Crackles

19

PRETEST ▶▶▶
1. What causes crackles?
2. Name the conditions associated with late inspiratory crackles.
3. Describe the characteristics of late inspiratory crackles.
4. What is the difference between coarse and fine crackles?
5. Name the conditions associated with early inspiratory and expiratory crackles.
6. Describe the characteristics of early inspiratory and expiratory crackles.
7. What are pleural crackles?
8. Describe the characteristics of pleural crackles.

Sound production

Crackles are short, explosive or popping sounds that are described according to their pitch, timing, and location. These characteristics change, depending on the underlying cause. Crackles are heard primarily through the chest wall with a stethoscope, but they may also be heard at the mouth with or without a stethoscope. Two different mechanisms are thought to generate crackles: air bubbling through secretions and the sudden, explosive opening of airways.

AIR BUBBLING THROUGH SECRETIONS
Air bubbling through secretions in the airways is widely accepted as one way crackles are produced. According to Dr. Paul Forgacs, this explanation is probably true when the trachea and mainstem bronchi are full of sputum, as commonly occurs with severe pulmonary edema and chronic bronchitis. However, Forgacs argues that the air bubbling theory doesn't explain why crackles are heard mainly during inspi-

ration and why they're heard over diseased lungs when sputum is usually absent. Further, in the smaller airways, airflow resistance, produced by the surface tension and viscosity of secretions, is too high to be overcome by the air pressure gradients that are present in healthy lungs during inspiration and expiration. Thus, another mechanism must explain how crackles are produced in the smaller airways.

SUDDEN, EXPLOSIVE OPENING OF AIRWAYS

At the end of expiration, peripheral airways in the lung bases close. At the beginning of inspiration, these peripheral airways remain closed, and inspired air flows to each lung apex first. The airways that are distal to the closed peripheral airways remain underexpanded until airway pressures and external forces, such as diaphragmatic movement and rib cage expansion, snap the airways open. The sudden opening of multiple collapsed peripheral airways and the associated explosive changes in air pressures are thought to produce the crackles heard over the lung bases in a healthy person who inhales deeply after a maximum exhalation. This mechanism is also thought to be responsible for the crackles heard in patients with atelectasis and interstitial lung disease.

Crackles associated with the abrupt equalization of pressures and the sudden opening of peripheral airways are characterized by a repetitive rhythm and loudness. These characteristics suggest that the airways open in the same sequence, at the same point in the respiratory cycle, and at the same approximate lung volumes. Crackles may be present in the lung bases of elderly people and occasionally in other healthy individuals. These crackles clear with coughing and have no pathological significance.

AGE ISSUE *Crackles aren't usually heard in infants; however, if present, they could indicate respiratory tract infection.*

Documentation

The timing of crackles is described as early inspiration, midinspiration, or late inspiration or early expiration, midexpiration, or late expiration. Pitch is usually described as high or low. Intensity varies and is described as loud or soft. Density is described as profuse or scanty. Duration, indicating the length of time that the crackles can be heard during inspiration or expiration, varies.

Late inspiratory crackles

Late inspiratory crackles are high-pitched, explosive sounds of variable intensity and density. They're heard mostly over dependent or poorly ventilated lung regions.

RELATED CONDITIONS

Conditions associated with late inspiratory crackles include atelectasis, resolving lobar pneumonia, interstitial fibrosis, and left-sided heart failure.

Atelectasis

Atelectasis, incomplete expansion of a lung area, is seen in postoperative and immobile patients and in patients with impaired diaphragmatic function. Prolonged shallow breathing, gravitational forces that close airways and deflate the lung bases, and mucus plugging the airways can all cause this condition. The result is poor ventilation in the affected areas and possible collapse of segmental or lobar bronchi. If small peripheral airways are involved, clinical findings may not be detectable. However, if a larger airway is involved, the clinical findings of decreased chest wall movement, a dull percussion note, and bronchial breath sounds are easily noted. Egophony, bronchophony, and whispered pectoriloquy typically are heard.

Crackles associated with atelectasis are produced by the sudden opening of collapsed small airways and adjoining alveoli. These crackles are high-pitched, explosive sounds heard late in inspiration. (**◆2-43**)

 AGE ISSUE *Premature infants may develop atelectasis due to a surfactant deficiency. Older adults on prolonged bedrest are especially susceptible to atelectasis due to changes in their musculoskeletal systems (such as kyphosis, which decreases lung expansion) and decreased respiratory and expiratory forces.*

Sound characteristics

In a postoperative patient who hasn't been coughing and deep breathing adequately, late inspiratory crackles are heard over the posterior bases of both lungs. This area is located between the eighth and tenth intercostal spaces from

Late inspiratory crackles heard over lower lobes

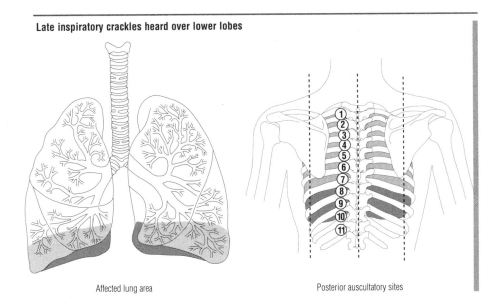

Affected lung area   Posterior auscultatory sites

the left posterior axillary line to the right posterior axillary line. The crackles begin late in inspiration and become more profuse toward the end of inspiration. They vary in intensity but are high pitched. (**◆2-44**)

Because crackles associated with atelectasis are poorly transmitted to the chest wall surface, their intensity and density change when the stethoscope is moved only a short distance. They aren't audible at the mouth. The patient's position also affects these crackles. For example, in an immobile patient, profuse crackles are heard in the dependent lung regions, but crackles are absent or scanty in nondependent lung regions. Also, crackles associated with atelectasis may clear somewhat with coughing.

Lobar pneumonia
In patients with resolving lobar pneumonia, crackles can be auscultated over lung areas where many alveoli are still filled with exudate, but the surrounding alveoli are aerated and have higher-than-normal ventilation. A large increase in air pressure gradients in the airways reaching these unaerated alveoli generates crackles as the airways are snapped open during late inspiration. These crackles have a sound similar to that of late inspiratory crackles heard over atelectatic areas, but they aren't affected by coughing or position changes. (**◆2-45**)

Late inspiratory crackles heard over right middle lobe

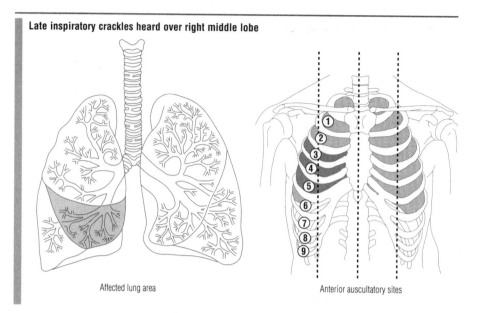

Affected lung area

Anterior auscultatory sites

AGE ISSUE *Older adults are at greater risk for developing pneumonia because their weakened chest muscles reduce their ability to clear secretions. Bacterial pneumonia is the most common type found in older adults; viral pneumonia, the second most common. Aspiration pneumonia occurs from impaired swallowing ability and a diminished gag reflex due to stroke or prolonged illness.*

Sound characteristics

In a patient with right middle lobe pneumonia, late inspiratory crackles are heard over the right anterior chest wall surface between the third and fifth intercostal spaces. They begin late in inspiration and become more profuse toward the end of inspiration. These crackles are typically high pitched. (◆2-46)

Interstitial fibrosis

Diffuse interstitial fibrosis impairs or destroys alveoli by filling them with abnormal cells or by scarring the lung tissue. Unaffected alveoli are usually hyperaerated. The lungs become stiff and more difficult to inflate, and airflow volumes usually decrease. Theoretically, the delayed opening of small airways during inspiration causes alveolar pressures in the affected lung tissue to fall more significantly than in healthy alveoli. This fall in alveolar pressures leads to an increased pressure gradient that generates repetitive

Auscultatory sites for late inspiratory crackles heard over lateral lung bases

Right lateral view

Left lateral view

late inspiratory crackles. (◆**2-47**) Coughing doesn't affect the profusion of crackles.

In mild interstitial fibrosis, crackles are auscultated at the end of inspiration over dependent lung regions — usually over the lateral lung bases when the patient is seated upright. Crackles can disappear if the patient inhales deeply, holds his breath, or leans forward. However, they usually recur when the patient returns to the upright position.

As interstitial fibrosis worsens, crackles are heard bilaterally over the posterior lung bases and spread upward toward the apices. In later stages of the disease, crackles aren't affected by position changes and may be heard throughout inspiration.

Either interstitial fibrosis or interstitial pneumonitis is associated with interstitial and alveolar infiltrates. Patients with a significant amount of lung involvement have dyspnea and a cough. Interstitial fibrosis may be caused by inhalation of heavy metals, the antibiotic nitrofurantoin, some chemotherapeutic agents, or prolonged inhalation of high concentrations of oxygen. In most cases, however, the cause is unknown.

Crackles heard in patients with known exposure to asbestos are thought to be an early sign of asbestosis, a lung disease characterized by pulmonary inflammation and fibrosis. The profusion of crackles increases in proportion to the amount of exposure to asbestos.

Lung area where right and left lateral late inspiratory crackles are heard

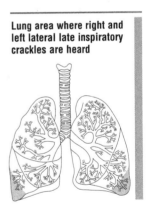

Pulmonary sarcoidosis, a noncaseating granulomatous disease of unknown cause that occurs in the lungs and other organs, may progress to interstitial fibrosis, resulting in restrictive lung disease. The connective tissue disorders rheumatoid arthritis and scleroderma are also associated with diffuse interstitial fibrosis.

Sound characteristics

Crackles associated with interstitial fibrosis caused by early asbestosis are initially heard in the midaxillary area over the lateral lung bases, the area over the seventh and eighth intercostal spaces. As the disease progresses, they may be heard over the posterior bases and may spread toward the apices. These crackles have a fine intensity and a short, discontinuous duration; they're heard during late inspiration. They have a high pitch heard that's best using the diaphragm of the stethoscope. (◆2-48)

Left-sided heart failure

Left-sided heart failure leads to fluid accumulation in the lung interstitium. Crackles are produced by the rapid equalization of pressures associated with the delayed opening of airways narrowed by pulmonary edema. They're commonly auscultated over the posterior lung bases in the early stages of pulmonary edema. Profuse and high pitched (◆2-49), these crackles are inaudible at the mouth. Common signs and symptoms of left-sided heart failure include dyspnea, orthopnea, hypoxia, cough, cyanosis or pallor, arrhythmias, and pulsus alternans. If pulmonary edema worsens, the airways are flooded with fluids. This results in severe hypoxia and low-pitched inspiratory and expiratory crackles that may be audible at the mouth and over the entire chest wall surface.

AUSCULTATION TIP *In the patient with heart failure, auscultate the lung bases first because crackles may not be audible after continued deep breaths.*

Sound characteristics

Crackles associated with early left-sided heart failure and pulmonary edema are heard bilaterally over the posterior lung bases, the area over the eighth, ninth, and tenth intercostal spaces. However, as pulmonary edema worsens, the crackles become more profuse and may be heard throughout the chest during late inspiration. Their intensity is fine in the early stages of left-sided heart failure but changes to

Late inspiratory crackles heard over posterior lung bases

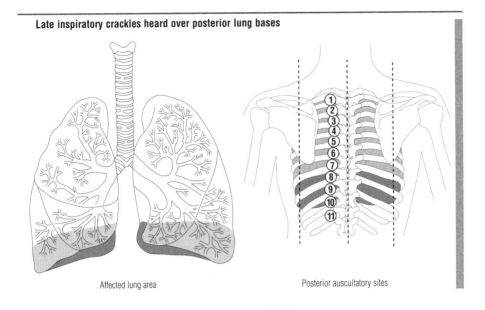

Affected lung area   Posterior auscultatory sites

a loud rattling sound as the patient's condition worsens; duration varies with the degree of failure. The crackles have a low pitch that's heard best using either the bell or the diaphragm of the stethoscope. (◆2-50)

Early inspiratory and expiratory crackles

Coarse crackles occur during early inspiration and during expiration. They're common in diffuse airway obstruction and may be heard over any chest wall area. Coarse crackles have a lower pitch than fine crackles and, although they're usually loud, may disappear after a cough; they're unaffected by the patient's position.

Coarse crackles typically are heard in an irregular rhythm that may be interrupted by short sequences of evenly spaced crackles having the same intensity. They're thought to be produced by the intermittent closure of large bronchi and the corresponding flow of a bolus of air through the obstructed area.

RELATED CONDITIONS
These crackles are associated with chronic bronchitis and bronchiectasis.

Auscultatory sites for early inspiratory crackles

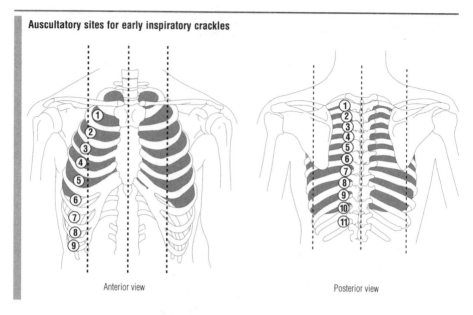

Anterior view

Posterior view

Lung area where early inspiratory crackles are heard

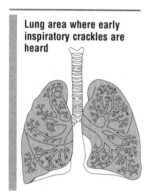

Chronic bronchitis

Early inspiratory crackles are associated with excessive mucus production in patients with chronic bronchitis. The proliferation and hypertrophy of mucous glands within the airways are caused by chronic exposure to airway irritants, such as cigarette smoke (the most common) and air pollution. Coughing, sputum production, and repeated respiratory tract infections are common in patients with chronic bronchitis. Chronic bronchitis can lead to chronic obstructive pulmonary disease. (◆2-51)

Sound characteristics

In patients with chronic bronchitis, crackles are heard early in inspiration, over all chest wall surfaces, and at the mouth. They're scanty and low pitched and aren't affected by the patient's position. (◆2-52)

Bronchiectasis

Bronchiectasis, irreversible dilation of the bronchi in selected lung segments, is characterized by chronic and copious production of yellow or green sputum and by fibrotic or atelectatic lung tissue surrounding the affected airways. Evidence of old inflammatory changes in the patient's lungs is usually noted on the chest X-ray.

Causes of bronchiectasis include foreign body obstruction, tumor, viral or bacterial pneumonia (particularly re-

Auscultatory sites for profuse early to midinspiratory crackles

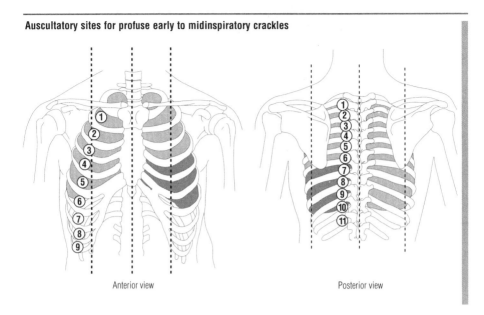

Anterior view   Posterior view

peated childhood pneumonias), chronic inflammatory or fibrotic lung disease, and tuberculosis. **(◆2-53)**

Sound characteristics
In patients with left midlung and lower lung area bronchiectasis, crackles are heard over the patient's anterior, posterior, and left lateral chest wall surfaces. The anterior area is located over the fourth, fifth, and sixth intercostal spaces, and the posterior area is located over the seventh, eighth, ninth, and tenth intercostal spaces. In patients with bronchiectasis, crackles tend to be profuse, low pitched, heard during early or midinspiration, and coarser than those associated with chronic bronchitis. Coughing may decrease the number of crackles heard, but changes in the patient's position don't affect them. **(◆2-54)**

Lung area where profuse early to midinspiratory crackles are heard

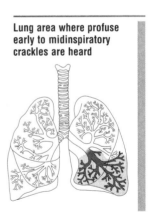

Pleural crackles

When the visceral and parietal pleural surfaces are damaged by fibrin deposits or inflammatory or neoplastic cells, they lose their ability to glide silently over each other during breathing, and their movements become jerky and periodically delayed. This motion produces loud, grating crackles known as *pleural crackles* or *pleural friction rub.* **(◆2-55)** These crackles may be present during the inspira-

Auscultatory sites for pleural crackles heard over right midlung

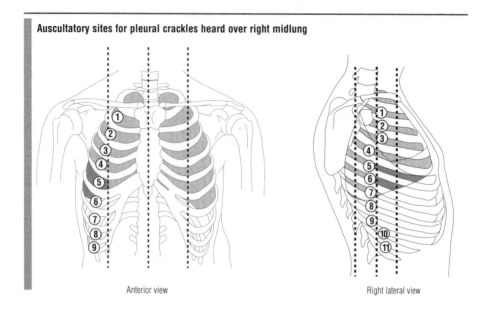

Anterior view

Right lateral view

tory and expiratory phases of the respiratory cycle, or they may be confined only to inspiration. Pleural crackles disappear if fluid accumulates between the pleurae.

Pleural crackles are usually associated with sharp pain during inspiration, which may cause the patient to spontaneously splint the affected side to minimize muscle movement and chest expansion. Asymmetrical chest wall movement in the patient may be palpable, and a rapid, shallow breathing pattern may also be present.

 AGE ISSUE *If a child has grunting respirations, reports chest pain, or protects his chest by holding it or lying on one side, auscultate for pleural crackles, an early sign of pleurisy.*

Lung area where right midlung pleural crackles are heard

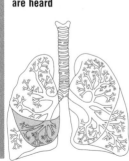

Sound characteristics

Pleural crackles are heard on the chest wall over the affected area. When pleural crackles are heard over the patient's right midlung — the area over the fifth and sixth intercostal spaces between the right midclavicular and right midaxillary lines — their intensity is loud and they have a coarse, grating quality and a low pitch. (◆2-56) Their duration is discontinuous. Pleural crackles may be heard during inspiration only or during both the inspiratory and expiratory phases of the respiratory cycle.

AUSCULTATION TIP *To distinguish between a pleural friction rub and a pericardial friction rub, have the patient hold his breath; if the rub continues, it's a pericardial friction rub.*

◀◀◀ POSTTEST

1. Define the term *crackles*.
2. Name two causes of crackles.
3. Explain how air bubbling through secretions produces crackles.
4. Explain how the sudden explosive opening of airways causes crackles.
5. What are the characteristics of crackles?
6. Name four conditions associated with late inspiratory crackles.
7. How does atelectasis generate crackles?
8. Describe the characteristics of crackles associated with atelectasis.
9. How are crackles generated in the patient with resolving lobar pneumonia?
10. Describe the characteristics of crackles associated with resolving lobar pneumonia.
11. How does interstitial fibrosis cause crackles?
12. How do the characteristics of crackles change as interstitial fibrosis worsens?
13. Describe the characteristics of crackles associated with interstitial fibrosis.
14. How are crackles generated in the patient with left-sided heart failure?
15. Describe the characteristics of crackles associated with left-sided heart failure.
16. How are coarse crackles thought to be produced?
17. Explain how early inspiratory crackles are generated in the patient with chronic bronchitis.
18. Describe the characteristics of crackles associated with chronic bronchitis.
19. What is bronchiectasis?
20. Describe the characteristics of crackles associated with bronchiectasis.
21. What causes pleural crackles?
22. Describe the characteristics of pleural crackles.

Adventitious sounds: Wheezes

20

PRETEST ▶▶▶
1. What causes wheezes?
2. What conditions are associated with wheezes?
3. Describe the following types of wheezes: expiratory polyphonic, fixed monophonic, sequential inspiratory, and random monophonic.
4. What type of wheeze is associated with asthma?
5. What type of wheeze is associated with widespread airflow obstruction?
6. What is stridor?

Sound production

Airflow pattern in narrowed bronchus

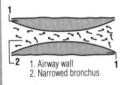

1. Airway wall
2. Narrowed bronchus

Oscillations of narrowed bronchus

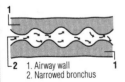

1. Airway wall
2. Narrowed bronchus

Wheezes — which are associated with such conditions as bronchospasm, airway thickening caused by mucosal swelling or muscle hypertrophy, inhalation of a foreign object, tumor, secretions, and dynamic airway compression — are musical sounds generated by air passing through a bronchus so narrowed as to be almost closed. The bronchus walls oscillate between closed and barely open positions; these oscillations generate audible sounds.

Wheezes are high-pitched, continuous sounds with frequencies of 200 Hz or greater and a duration of 250 msec or more. Their duration is long enough to carry an audible pitch similar to a musical tone. Wheezes are described by their timing within the respiratory cycle; they may be heard during inspiration or expiration or continuously throughout the respiratory cycle. Wheezes can be further described as localized or diffuse, episodic or chronic.

FREQUENCY

A wheeze's pitch is determined by the emitted sound's frequency, which can vary widely over a five-octave range. These differences in frequency are attributed to the airway size and elasticity and to airflow rates through the narrowed bronchus. Theoretically, large, flabby, narrowed airways generate low-pitched sounds; stiff, narrowed airways generate high-pitched sounds. Pitch isn't believed to be related to airway length or width or to air density.

Wheezes of different pitches may occur simultaneously or may overlap, and the pitch of a single wheeze may change during inspiration and expiration. *Monophonic wheezes* produce single musical notes that may vary in duration and may overlap. *Polyphonic wheezes* combine several distinct musical tones.

TRANSMISSION

Wheezes are transmitted better through airways than through lung tissue, which absorbs high-frequency sounds. Wheezes that sound louder over central airways than over peripheral lung tissue may be sounds transmitted through the central airways but not produced by them. Wheezes heard at the mouth in severe airway obstruction may not be audible during auscultation of the chest wall surfaces.

 ALERT *Be alert for impending respiratory failure if wheezes are audible at the mouth and breath sounds are diminished or absent.*

Expiratory polyphonic wheezes

Polyphonic wheezes are believed to be caused by the dynamic compression of large airways during expiration. In healthy individuals, polyphonic wheezes can sometimes be auscultated during a maximal forced expiration when the dynamic compression occurs simultaneously in all airways. No regional variations exist in airway resistance or in lung compliance; consequently, polyphonic wheezes are heard throughout the lungs during expiration.

Polyphonic wheezes are loud and widely transmitted; when heard during regular breathing, they indicate widespread airflow obstruction.

In patients with widespread airflow obstruction (as in asthma and chronic bronchitis), elastic recoil properties, peripheral airway resistance, and airway mechanics are altered

Auscultatory sites for expiratory polyphonic wheezes

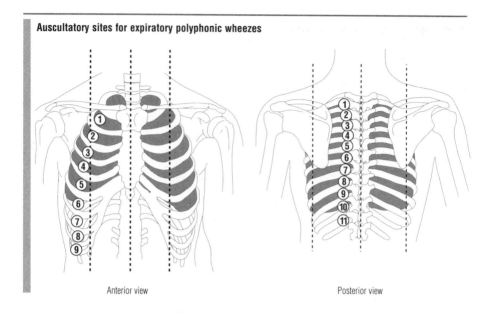

Anterior view   Posterior view

Lung area where expiratory polyphonic wheezes are heard

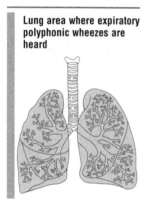

throughout both lungs. Together, these abnormal changes affect the timing of dynamic airway compression within the bronchi, which alters normal sound production. For example, during sequential forceful expirations, the central bronchi are compressed first when elastic recoil of the airways is low or when peripheral airway resistance is high. This compression produces a series of sounds, beginning with a monophonic wheeze and quickly followed by bitonal sounds. Soon, the full complement of sounds that make up polyphonic wheezes are audible through the stethoscope. These multiple musical tones begin simultaneously and maintain a constant pitch that rises sharply at the end of expiration. (◆2-57)

Sound characteristics
Expiratory polyphonic wheezes are usually heard over the anterior, posterior, and lateral chest wall surfaces during expiration. Their intensity is usually described as loud and musical, and their duration is continuous. They have a high pitch that's heard best using the diaphragm of the stethoscope. (◆2-58)

Auscultatory sites for fixed monophonic wheezes

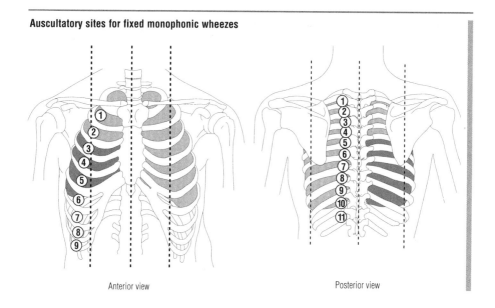

Anterior view   Posterior view

Fixed monophonic wheezes

Fixed monophonic wheezes, which have a constant pitch and a single musical tone, are generated by the oscillations of a large, partially obstructed bronchus. The obstruction may be caused by a tumor, a foreign body, bronchial stenosis, or an intrabronchial granuloma; each of these conditions narrows the airway but doesn't slow airflow. Fixed monophonic wheezes are low pitched and are transmitted throughout the lungs. (◆2-59)

In a patient with bronchial stenosis, the pitch of inspiratory and expiratory monophonic wheezes varies, depending on airway rigidity. Monophonic wheezes may disappear when the patient is in a supine position or turns from side to side and may clear slightly following coughing.

Lung area where fixed monophonic wheezes are heard

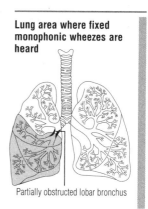

Partially obstructed lobar bronchus

Sound characteristics
In a patient with a partially obstructed right lobar bronchus, fixed monophonic wheezes are heard over the anterior, posterior, and right lateral chest wall surfaces. The anterior area is located over the third, fourth, fifth, and sixth intercostal spaces; the posterior area is located over the fifth, sixth, seventh, eighth, ninth, and tenth intercostal spaces. Fixed monophonic wheezes are usually loud, and their duration is continuous. They may be heard during inspiration,

Lung area where sequential inspiratory wheezes are heard	Auscultatory sites for sequential inspiratory wheezes

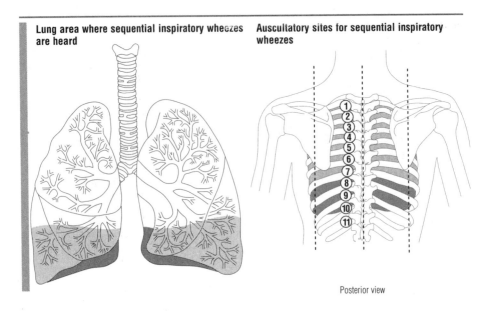

Posterior view

expiration, or throughout the respiratory cycle. They have a low pitch that's heard best using the diaphragm of the stethoscope. (◆2-60)

Sequential inspiratory wheezes

Sequential inspiratory wheezes are sometimes heard over the lung bases in patients with interstitial fibrosis, asbestosis, or fibrosing alveolitis. These monophonic wheezes are generated by airways that open late in inspiration in unaerated lung regions. The rapid inflow of air precipitates airway wall vibrations, generating sequential inspiratory wheezes. A single, short inspiratory wheeze or a brief sequence of monophonic inspiratory wheezes with different pitches may be heard along with the crackles normally associated with interstitial fibrosis. (◆2-61)

Sound characteristics
Sequential inspiratory wheezes are usually heard over the lateral and posterior lung bases. On the posterior chest wall surface, this area is located over the eighth, ninth, and tenth intercostal spaces; on the lateral chest wall surfaces, it's located over the seventh and eighth intercostal spaces. Sequential inspiratory wheezes have a loud intensity and a continuous duration. They occur throughout inspiration but

Auscultatory sites for sequential inspiratory wheezes *(continued)*

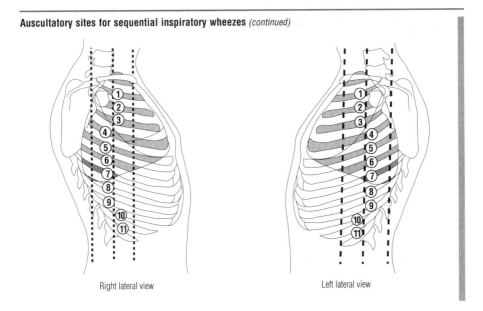

Right lateral view   Left lateral view

are more predominant in late inspiration. They have a high pitch that's heard best using the diaphragm of the stethoscope. (◆2-62)

Random monophonic wheezes

Airways narrowed by bronchospasm or mucosal swelling produce single or multiple monophonic wheezes that may occur during inspiration, expiration, or throughout the respiratory cycle. Multiple wheezes can occur randomly and vary in duration. Random monophonic wheezes produced in the large central airways are loud and widely transmitted throughout the lung (see *Lung area where random monophonic wheezes are heard,* page 182); they can be heard at a distance from the mouth. (◆2-63) In contrast, random monophonic wheezes produced in the peripheral airways are weaker and are filtered as they're transmitted to the chest wall; these sounds are audible only over the chest wall.

Random monophonic wheezes are typically heard in patients with severe status asthmaticus. In such patients, airway resistance is high; dynamic airway compression moves from the central airways toward the smaller peripheral airways, where the airflow rate is too low to produce airway wall vibrations or sounds.

Lung area where random monophonic wheezes are heard

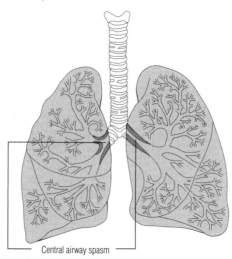

Central airway spasm

The progressive airway obstruction that occurs in patients with status asthmaticus can cause a dynamic pattern of wheezing. First, monophonic wheezes are heard only during expiration; then they're heard throughout the respiratory cycle. Monophonic wheezes may also be heard at the mouth because the high-frequency sounds are transmitted through larger airways.

ALERT *As status asthmaticus becomes more severe, all wheezes heard over the chest wall surfaces disappear because a combination of air trapping and severe airway narrowing causes the site of dynamic airway compression to move toward the lung periphery; this phenomenon is called a* silent chest. *Silent chest is commonly accompanied by hypercapnia (increased carbon dioxide levels in the blood) and acidosis, both of which are life-threatening.*

Sound characteristics
Random monophonic wheezes are usually heard over the anterior, posterior, and lateral chest wall surfaces. They're usually loud, and their duration is continuous. These wheezes occur throughout the respiratory cycle, and expiration is frequently prolonged. They have a high pitch that's heard best using the diaphragm of the stethoscope. (◆2-64)

Auscultatory sites for random monophonic wheezes

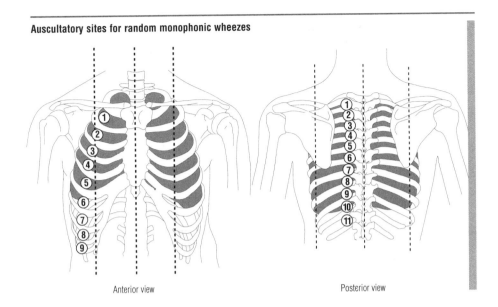

Anterior view   Posterior view

Stridor

Stridor is a loud musical sound that's heard at a distance from the patient, usually without a stethoscope. It's caused by laryngeal spasm and mucosal swelling, which contract the vocal cords and narrow the airway. The loud monophonic wheeze of stridor is typically heard during inspiration, but it also may be heard throughout the respiratory cycle as airway constriction increases. Its intensity distinguishes it from other monophonic wheezes. (◆2-65) Stridor is associated with whooping cough, laryngeal tumors, tracheal stenosis, aspiration of a foreign object and, most commonly, severe upper respiratory tract infections.

ALERT *Children with acute epiglottitis have stridor that worsens when in the supine position. During throat examination, trained personnel, such as an anesthesiologist, should be present to secure an emergency airway. Special equipment, such as a laryngoscope and an endotracheal tube, should be available because a tongue blade can cause sudden, complete airway obstruction.*

Sound characteristics

Severe stridor is heard without a stethoscope; however, in a patient with a less-pronounced laryngeal spasm, stridor may be heard by auscultating over the larynx. Stridor is loud and has a continuous duration. It may be heard during inspiration or throughout the respiratory cycle. Its high pitch resembles a crowing sound. (**◆2-66**)

ALERT *If the patient with stridor is drooling, the epiglottis may be severely swollen. Don't examine the mouth. Any stimulus could worsen the airway occlusion. Immediate medical intervention is necessary. The patient may require intubation or emergency tracheotomy.*

◀◀◀ POSTTEST

1. Define the term *wheezes*.
2. What are the characteristics of wheezes?
3. How is a wheeze's pitch determined?
4. How are wheezes transmitted?
5. Name five conditions associated with wheezes.
6. How are expiratory polyphonic wheezes generated?
7. How does widespread airflow obstruction change the sounds heard during auscultation?
8. Describe the characteristics of expiratory polyphonic wheezes.
9. How are fixed monophonic wheezes generated?
10. Describe the characteristics of fixed monophonic wheezes.
11. What conditions are associated with sequential inspiratory wheezes?
12. How are sequential inspiratory wheezes generated?
13. Describe the characteristics of sequential inspiratory wheezes.
14. How are random monophonic wheezes generated?
15. What condition is associated with random monophonic wheezes?
16. How can the characteristics of random monophonic wheezes be expected to change as asthma worsens?
17. Describe the characteristics of random monophonic wheezes.
18. What is stridor?
19. What causes stridor?
20. Describe stridor's characteristics.

Auscultation findings for common disorders

The following is a compilation of various cardiac and pulmonary disorders along with associated auscultation findings. The patient may not present with every assessment finding listed for each disorder.

Disorder	Abnormal breath sounds	Abnormal heart sounds
Aortic prosthetic valve		▪ Systolic murmur ▪ Aortic opening click ▪ Aortic closing click ▪ Interval between aortic closing click and P_2 that widens during inspiration
Aortic regurgitation		▪ Early diastolic murmur often with midsystolic murmur ▪ "Cooing dove" or "musical" diastolic murmur signifies rupture or retroversion of an aortic cusp ▪ Diminished A_2 ▪ Early ejection click ▪ Paradoxical S_2 split ▪ S_3 and S_4 ▪ Soft S_1
Aortic valvular stenosis	▪ Inspiratory and expiratory crackles over the posterior lung bases associated with heart failure	▪ Midsystolic murmur ▪ Paradoxical S_2 split ▪ Delayed A_2 and shortened A_2–P_2 interval ▪ Aortic ejection sound radiates widely to neck and along great vessels ▪ Systolic ejection sound if not severely stenotic
Asbestosis	▪ High-pitched crackles heard at the end of inspiration ▪ Pleural friction rub	Only when severe: ▪ Paradoxical S_2 split ▪ Right ventricular heave

Disorder	Abnormal breath sounds	Abnormal heart sounds
Asthma	• Diminished breath sounds • Musical, high-pitched expiratory polyphonic wheezes • With status asthmaticus, loud and continuous random monophonic wheezes heard, along with prolonged expiration and possible silent chest if severe	
Atelectasis	• High-pitched, hollow, tubular bronchial breath sounds, crackles, wheezes • Fine, high-pitched late inspiratory crackles • Bronchophony • Egophony • Whispered pectoriloquy • I:E ratio: I > E over empty lung field	
Atrial septal defect		• Wide, fixed S_2 split • Pulmonic ejection sound • Tricuspid component louder than mitral component
Bronchial stenosis	• Loud, continuous, low-pitched, fixed, monophonic wheezes that may disappear when in supine position or turning side to side	
Bronchiectasis	• Profuse, low-pitched crackles heard during midinspiration	
Chronic bronchitis	• Scanty, low-pitched, early inspiratory crackles not affected by patient's position • Loud, musical, high-pitched, expiratory polyphonic wheezes	

Disorder	Abnormal breath sounds	Abnormal heart sounds
Chronic obstructive pulmonary disease	• Diminished, low-pitched breath sounds • Sonorous and/or sibilant wheezes • Inaudible bronchophony, egophony, and whispered pectoriloquy • Prolonged expiration • Fine inspiratory crackles	• Paradoxical S_2 split • Right ventricular heave
Fibrosing alveolitis	• Loud, continuous, high-pitched, sequential wheezes	
Hypertrophic obstructive cardiomyopathy	• Inspiratory and expiratory crackles over the posterior lung bases associated with heart failure	• Midsystolic murmur that becomes louder with Valsalva's maneuver • Paradoxical S_2 split • S_3 and S_4
Interstitial pulmonary fibrosis	• Late inspiratory fine crackles not affected by coughing but may disappear with position change, deep inhalation, or holding of breath • High-pitched bronchial or bronchovesicular breath sounds heard over lower lung regions that are audible through inspiration and expiration • Loud, high-pitched sequential wheezes of continuous duration • Whispered pectoriloquy	
Laryngeal spasm	• Stridor heard during inspiration	
Malfunctioning aortic prosthetic valve	• Inspiratory and expiratory crackles over the posterior lung bases associated with heart failure	• Long systolic ejection murmur • Absent or softened aortic opening click • Absent or diminished aortic closing click • Diastolic murmur

Disorder	Abnormal breath sounds	Abnormal heart sounds
Malfunctioning mitral prosthetic valve	▪ Inspiratory and expiratory crackles over the posterior lung bases associated with heart failure	▪ Holosystolic murmur ▪ New diastolic murmur or change in intensity or duration of existing one ▪ Diastolic rumble ▪ Absent mitral closing click
Mitral prosthetic valve		▪ Systolic murmur ▪ Mitral closing click ▪ Mitral opening click depending on valve type ▪ Diastolic murmur depending on valve type
Mitral regurgitation	▪ Inspiratory and expiratory crackles over the posterior lung bases associated with heart failure	▪ Holosystolic murmur ▪ Decreased S_1 intensity ▪ Increased S_2 intensity ▪ Accented P_2 ▪ S_3 and S_4 ▪ Persistent A_2–P_2 splitting during expiration
Mitral stenosis		▪ With moderate stenosis, early and late diastolic murmur ▪ In severe stenosis, possibly holo-diastolic murmur ▪ Loud M_1 ▪ Intensified S_1 except with a calcified valve, which produces a soft S_1 ▪ Split S_2 ▪ Opening snap except with a calcified valve ▪ S_3
Mitral valve prolapse		▪ Late systolic or holosystolic murmur ▪ Nonejection midsystolic click that's variable in intensity and timing ▪ Precordial honk
Myocardial infarction	▪ Inspiratory and expiratory crackles over the posterior lung bases associated with heart failure	▪ S_3 and S_4 ▪ S_1 and S_2 faint and poor quality ▪ Paradoxical splitting S_2 with left ventricular dysfunction or left bundle-branch block ▪ Physiologic splitting S_2 with right bundle-branch block, ventricular septal defect, severe mitral regurgitation

Disorder	Abnormal breath sounds	Abnormal heart sounds
Myocardial infarction *(continued)*		▪ Harsh holosystolic murmur, crescendo-decrescendo with palpable thrill if ventricular septal rupture
Patent ductus arteriosus		▪ Continuous murmur reaching maximum intensity during late systole ▪ Murmur envelops S_2 ▪ Paradoxical S_2 split ▪ S_3 ▪ Diastolic flow rumble
Pericarditis		▪ Pericardial friction rub, which has both a systolic (loudest) and diastolic component ▪ Scratchy, superficial quality
Pleural effusion	▪ Absent or diminished low-pitched breath sounds ▪ Occasionally loud bronchial breath sounds ▪ Normal breath sounds on contralateral side ▪ Bronchophony, egophony, and whispered pectoriloquy at upper border of pleural effusion	
Pneumonia	▪ High-pitched, tubular bronchial breath sounds heard over affected area during inspiration and expiration ▪ Bronchophony ▪ Egophony ▪ Whispered pectoriloquy ▪ Late inspiratory crackles not affected by coughing or position changes ▪ I:E ratio 1:1	
Pneumothorax	▪ Absent or diminished low-pitched breath sounds ▪ Inaudible bronchophony, egophony, and whispered pectoriloquy ▪ Normal breath sounds heard on contralateral side	

Disorder	Abnormal breath sounds	Abnormal heart sounds
Pulmonary edema	• Inspiratory and expiratory crackles over the posterior lung bases; as pulmonary edema worsens, crackles more profuse and heard during late inspiration • Wheezes	• S_3 or S_4
Pulmonic regurgitation		• Early diastolic murmur intensified during inspiration • Loud P_2
Pulmonic valve stenosis		• Midsystolic murmur • Systolic ejection click • P_2 absent or diminished and delayed • Normal or widened S_2 split • S_4
Supravalvular aortic stenosis		• Midsystolic murmur • Normal S_2 split • No aortic ejection sound
Tricuspid stenosis		• Mid- to late-diastolic murmur, which increases during inspiration and fades during expiration • Normal S_2 or may be split during inspiration • Opening snap seldom heard
Ventricular septal defect		• Loud holosystolic murmur • Persistent A_2–P_2 splitting during expiration • S_3 followed by short low-frequency diastolic murmur • Pulmonic ejection sound • Loud P_2
Whooping cough	• Stridor	

Assessment flowcharts for common auscultation findings

These flowcharts will help you quickly recognize and provide interventions for seven common and, possibly, life-threatening findings. Each flowchart includes presenting signs and symptoms, diagnostic tests, treatment, and follow-up care.

FLOWCHART ABBREVIATIONS

ABC	Airway, breathing, and circulation
ABG	Arterial blood gas
ACE	Angiotensin-converting enzyme
ARDS	Acute respiratory distress syndrome
AV	Atrioventricular
CBC	Complete blood count
CT	Computed tomography
CXR	Chest X-ray
DX	Diagnosis
ECG	Electrocardiogram
F/U	Follow-up
GERD	Gastroesophageal reflux disease
HEENT	Head, ears, eyes, nose, and throat
HPI	History of present illness
IABP	Intra-aortic balloon pump
I.V.	Intravenous
JVD	Jugular vein distention
MI	Myocardial infarction
MRI	Magnetic resonance imaging
PCI	Percutaneous coronary intervention
PE	Physical examination
PFT	Pulmonary function tests
SLE	Systemic lupus erythematous
SOB	Shortness of breath
Sub-Q	Subcutaneous
Tc	Technetium
TX	Treatment
$\dot{V}/\dot{Q}$	Ventilation-perfusion

GALLOP, ATRIAL OR VENTRICULAR

An atrial gallop (S_4) is auscultated immediately before the first heart sound (S_1) and results from abnormal forceful atrial contraction caused by augmented ventricular filling or by decreased left ventricular compliance. A ventricular gallop (S_3) is auscultated immediately after the second heart sound (S_2) and results from rapid deceleration of blood entering a stiff, noncompliant ventricle or rapid acceleration of blood associated with increased flow into the ventricle. These two left ventricular sounds may be heard over the mitral area.

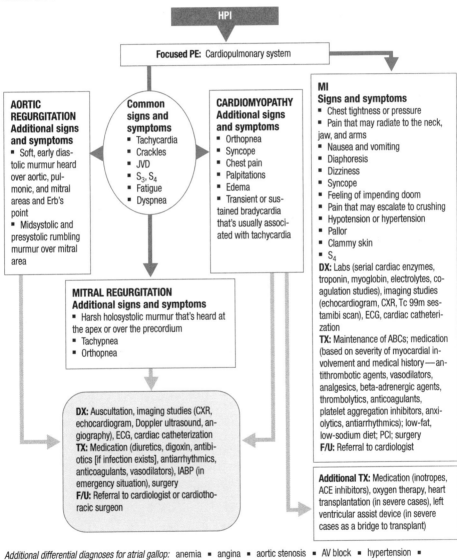

Additional differential diagnoses for atrial gallop: anemia ▪ angina ▪ aortic stenosis ▪ AV block ▪ hypertension ▪ pulmonary embolism

Additional differential diagnoses for ventricular gallop: heart failure ▪ thyrotoxicosis

MURMUR

Murmurs are auscultatory sounds heard within the heart chambers or major arteries. They're classified by their timing and duration in the cardiac cycle, auscultatory location, loudness, configuration, pitch, and quality.

Murmurs can reflect accelerated blood flow through a normal or an abnormal valve; forward blood flow through a narrowed or irregular valve or into a dilated vessel; blood backflow through an incompetent valve, septal defect, or patent ductus arteriosus; or decreased blood viscosity. Typically the result of organic heart disease, murmurs occasionally signal an emergency. Murmurs may also result from surgical implantation of a prosthetic valve.

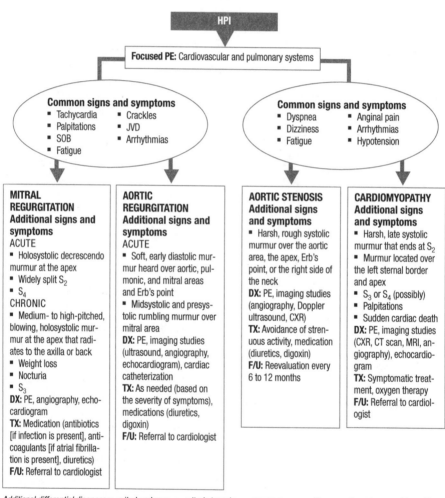

Additional differential diagnoses: mitral prolapse ▪ mitral stenosis ▪ myxomas ▪ papillary muscle rupture ▪ tricuspid regurgitation ▪ tricuspid stenosis

Other causes: prosthetic valve replacement

CRACKLES

A common finding in certain cardiovascular and pulmonary diseases, crackles are characterized by short, explosive, or popping sounds usually heard during inspiration. They may be described as coarse (loud and low in pitch) or fine (less intense and high in pitch). Crackles are believed to result either when air bubbles through secretions in the airways or when the airways open suddenly and explosively.

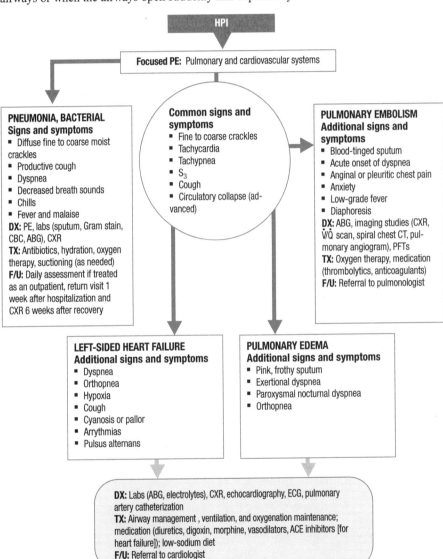

HPI

Focused PE: Pulmonary and cardiovascular systems

PNEUMONIA, BACTERIAL
Signs and symptoms
- Diffuse fine to coarse moist crackles
- Productive cough
- Dyspnea
- Decreased breath sounds
- Chills
- Fever and malaise

DX: PE, labs (sputum, Gram stain, CBC, ABG), CXR
TX: Antibiotics, hydration, oxygen therapy, suctioning (as needed)
F/U: Daily assessment if treated as an outpatient, return visit 1 week after hospitalization and CXR 6 weeks after recovery

Common signs and symptoms
- Fine to coarse crackles
- Tachycardia
- Tachypnea
- S_3
- Cough
- Circulatory collapse (advanced)

PULMONARY EMBOLISM
Additional signs and symptoms
- Blood-tinged sputum
- Acute onset of dyspnea
- Anginal or pleuritic chest pain
- Anxiety
- Low-grade fever
- Diaphoresis

DX: ABG, imaging studies (CXR, V̇/Q̇ scan, spiral chest CT, pulmonary angiogram), PFTs
TX: Oxygen therapy, medication (thrombolytics, anticoagulants)
F/U: Referral to pulmonologist

LEFT-SIDED HEART FAILURE
Additional signs and symptoms
- Dyspnea
- Orthopnea
- Hypoxia
- Cough
- Cyanosis or pallor
- Arrythmias
- Pulsus alternans

PULMONARY EDEMA
Additional signs and symptoms
- Pink, frothy sputum
- Exertional dyspnea
- Paroxysmal nocturnal dyspnea
- Orthopnea

DX: Labs (ABG, electrolytes), CXR, echocardiography, ECG, pulmonary artery catheterization
TX: Airway management , ventilation, and oxygenation maintenance; medication (diuretics, digoxin, morphine, vasodilators, ACE inhibitors [for heart failure]); low-sodium diet
F/U: Referral to cardiologist

Additional differential diagnoses: ARDS ▪ asthma (acute) ▪ bronchiectasis ▪ bronchitis (chronic) ▪ interstitial fibrosis of the lungs ▪ legionnaires' disease ▪ lung abscess ▪ pneumonia ▪ psittacosis ▪ pulmonary tuberculosis ▪ sarcoidosis ▪ silicosis ▪ tracheobronchitis

PLEURAL FRICTION RUB (PLEURAL CRACKLES)

Commonly resulting from a pulmonary disorder or trauma, pleural friction rub (or *pleural crackles*) is a coarse, grating sound that may be auscultated over one or both lungs during inspiration or expiration. Pleural friction rub indicates inflammation of the visceral and parietal pleural lining. The resultant fibrinous exudate causes the pleural surfaces to lose their ability to glide silently over each other during breathing. Patients with pleural friction rub typically have sharp inspiratory pain, causing them to splint the affected side in an attempt to reduce muscle and chest wall movement.

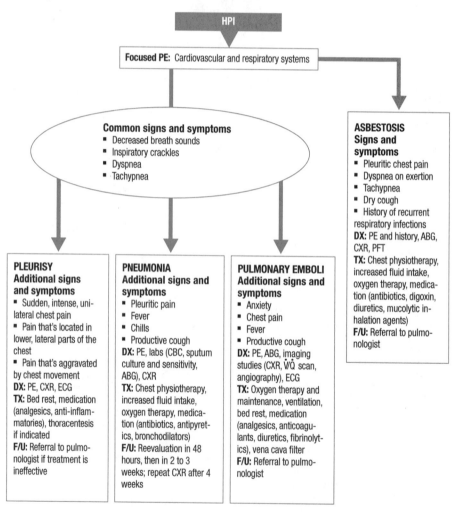

HPI

Focused PE: Cardiovascular and respiratory systems

Common signs and symptoms
- Decreased breath sounds
- Inspiratory crackles
- Dyspnea
- Tachypnea

ASBESTOSIS
Signs and symptoms
- Pleuritic chest pain
- Dyspnea on exertion
- Tachypnea
- Dry cough
- History of recurrent respiratory infections

DX: PE and history, ABG, CXR, PFT
TX: Chest physiotherapy, increased fluid intake, oxygen therapy, medication (antibiotics, digoxin, diuretics, mucolytic inhalation agents)
F/U: Referral to pulmonologist

PLEURISY
Additional signs and symptoms
- Sudden, intense, unilateral chest pain
- Pain that's located in lower, lateral parts of the chest
- Pain that's aggravated by chest movement

DX: PE, CXR, ECG
TX: Bed rest, medication (analgesics, anti-inflammatories), thoracentesis if indicated
F/U: Referral to pulmonologist if treatment is ineffective

PNEUMONIA
Additional signs and symptoms
- Pleuritic pain
- Fever
- Chills
- Productive cough

DX: PE, labs (CBC, sputum culture and sensitivity, ABG), CXR
TX: Chest physiotherapy, increased fluid intake, oxygen therapy, medication (antibiotics, antipyretics, bronchodilators)
F/U: Reevaluation in 48 hours, then in 2 to 3 weeks; repeat CXR after 4 weeks

PULMONARY EMBOLI
Additional signs and symptoms
- Anxiety
- Chest pain
- Fever
- Productive cough

DX: PE, ABG, imaging studies (CXR, V̇/Q̇ scan, angiography), ECG
TX: Oxygen therapy and maintenance, ventilation, bed rest, medication (analgesics, anticoagulants, diuretics, fibrinolytics), vena cava filter
F/U: Referral to pulmonologist

Additional differential diagnoses: lung cancer ▪ rheumatoid arthritis ▪ SLE ▪ tuberculosis (pulmonary)

Other causes: radiation therapy ▪ thoracic surgery

LOW-PITCHED WHEEZES (RHONCHI)

Low-pitched wheezes, previously known as *sonorous rales* or *sonorous rhonchi,* are continuous and low-pitched, producing a snoring sound. The term *rhonchi* may still be used in the clinical setting to describe a rough, rumbling low-pitched sound heard mainly on expiration. Low-pitched wheezes are typically produced when fluid or secretions narrow the large airways, and often change in sound or disappear with coughing.

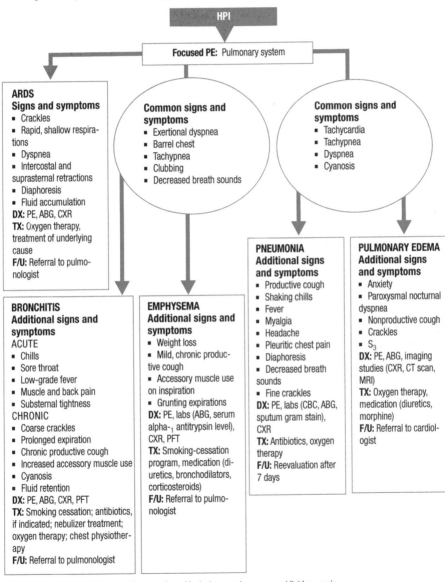

HPI

Focused PE: Pulmonary system

ARDS
Signs and symptoms
- Crackles
- Rapid, shallow respirations
- Dyspnea
- Intercostal and suprasternal retractions
- Diaphoresis
- Fluid accumulation
DX: PE, ABG, CXR
TX: Oxygen therapy, treatment of underlying cause
F/U: Referral to pulmonologist

Common signs and symptoms
- Exertional dyspnea
- Barrel chest
- Tachypnea
- Clubbing
- Decreased breath sounds

Common signs and symptoms
- Tachycardia
- Tachypnea
- Dyspnea
- Cyanosis

BRONCHITIS
Additional signs and symptoms
ACUTE
- Chills
- Sore throat
- Low-grade fever
- Muscle and back pain
- Substernal tightness
CHRONIC
- Coarse crackles
- Prolonged expiration
- Chronic productive cough
- Increased accessory muscle use
- Cyanosis
- Fluid retention
DX: PE, ABG, CXR, PFT
TX: Smoking cessation; antibiotics, if indicated; nebulizer treatment; oxygen therapy; chest physiotherapy
F/U: Referral to pulmonologist

EMPHYSEMA
Additional signs and symptoms
- Weight loss
- Mild, chronic productive cough
- Accessory muscle use on inspiration
- Grunting expirations
DX: PE, labs (ABG, serum alpha-$_1$ antitrypsin level), CXR, PFT
TX: Smoking-cessation program, medication (diuretics, bronchodilators, corticosteroids)
F/U: Referral to pulmonologist

PNEUMONIA
Additional signs and symptoms
- Productive cough
- Shaking chills
- Fever
- Myalgia
- Headache
- Pleuritic chest pain
- Diaphoresis
- Decreased breath sounds
- Fine crackles
DX: PE, labs (CBC, ABG, sputum gram stain), CXR
TX: Antibiotics, oxygen therapy
F/U: Reevaluation after 7 days

PULMONARY EDEMA
Additional signs and symptoms
- Anxiety
- Paroxysmal nocturnal dyspnea
- Nonproductive cough
- Crackles
- S_3
DX: PE, ABG, imaging studies (CXR, CT scan, MRI)
TX: Oxygen therapy, medication (diuretics, morphine)
F/U: Referral to cardiologist

Additional differential diagnoses: asthma ▪ bronchiectasis ▪ pulmonary coccidioidomycosis

Other causes: bronchoscopy ▪ foreign body aspiration ▪ PFTs ▪ respiratory therapy

STRIDOR

A loud, musical respiratory sound, stridor results from an obstruction in the trachea or larynx and may be heard without a stethoscope. Usually heard during inspiration, this sign may also occur during expiration in severe upper-airway obstruction. Stridor may signal a life-threatening condition requiring prompt emergency interventions.

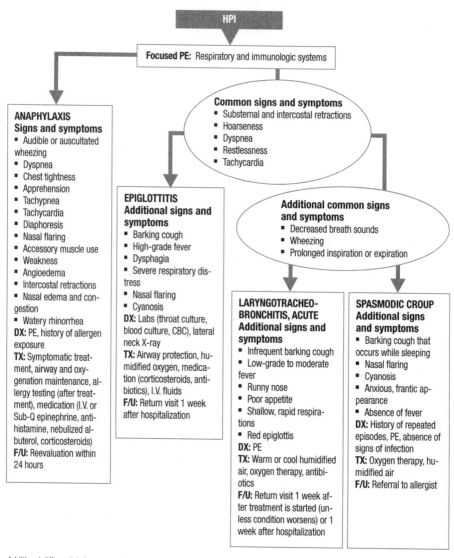

HPI

Focused PE: Respiratory and immunologic systems

Common signs and symptoms
- Substernal and intercostal retractions
- Hoarseness
- Dyspnea
- Restlessness
- Tachycardia

ANAPHYLAXIS
Signs and symptoms
- Audible or auscultated wheezing
- Dyspnea
- Chest tightness
- Apprehension
- Tachypnea
- Tachycardia
- Diaphoresis
- Nasal flaring
- Accessory muscle use
- Weakness
- Angioedema
- Intercostal retractions
- Nasal edema and congestion
- Watery rhinorrhea
DX: PE, history of allergen exposure
TX: Symptomatic treatment, airway and oxygenation maintenance, allergy testing (after treatment), medication (I.V. or Sub-Q epinephrine, antihistamine, nebulized albuterol, corticosteroids)
F/U: Reevaluation within 24 hours

EPIGLOTTITIS
Additional signs and symptoms
- Barking cough
- High-grade fever
- Dysphagia
- Severe respiratory distress
- Nasal flaring
- Cyanosis
DX: Labs (throat culture, blood culture, CBC), lateral neck X-ray
TX: Airway protection, humidified oxygen, medication (corticosteroids, antibiotics), I.V. fluids
F/U: Return visit 1 week after hospitalization

Additional common signs and symptoms
- Decreased breath sounds
- Wheezing
- Prolonged inspiration or expiration

LARYNGOTRACHEO-BRONCHITIS, ACUTE
Additional signs and symptoms
- Infrequent barking cough
- Low-grade to moderate fever
- Runny nose
- Poor appetite
- Shallow, rapid respirations
- Red epiglottis
DX: PE
TX: Warm or cool humidified air, oxygen therapy, antibiotics
F/U: Return visit 1 week after treatment is started (unless condition worsens) or 1 week after hospitalization

SPASMODIC CROUP
Additional signs and symptoms
- Barking cough that occurs while sleeping
- Nasal flaring
- Cyanosis
- Anxious, frantic appearance
- Absence of fever
DX: History of repeated episodes, PE, absence of signs of infection
TX: Oxygen therapy, humidified air
F/U: Referral to allergist

Additional differential diagnoses: airway trauma ▪ hypocalcemia ▪ inhalation injury ▪ laryngeal tumor ▪ mediastinal tumor ▪ retrosternal thyroid ▪ thoracic aortic aneurysm

Other causes: bronchoscopy ▪ foreign body aspiration ▪ laryngoscopy ▪ neck surgery ▪ prolonged intubation

WHEEZING

Wheezes are high-pitched, continuous musical sounds that result when air passes rapidly through a narrowed bronchus that oscillates between being barely open to being completely closed. Causes of airway narrowing include bronchospasm; obstructive lung disease; mucosal thickening or edema; partial obstruction from a tumor, foreign body, or secretions; use of a beta-adrenergic blocker; and extrinsic pressure, as in tension pneumothorax or goiter. Wheezing may signal a medical emergency requiring immediate medical intervention.

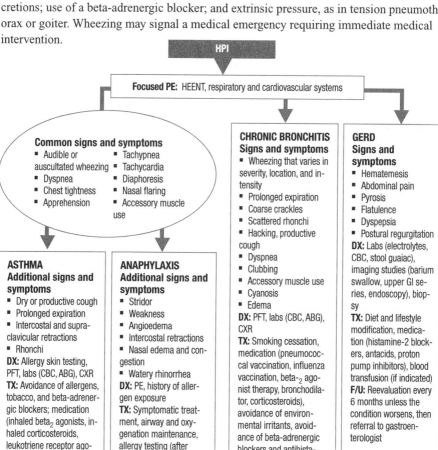

HPI

Focused PE: HEENT, respiratory and cardiovascular systems

Common signs and symptoms
- Audible or auscultated wheezing
- Dyspnea
- Chest tightness
- Apprehension
- Tachypnea
- Tachycardia
- Diaphoresis
- Nasal flaring
- Accessory muscle use

ASTHMA
Additional signs and symptoms
- Dry or productive cough
- Prolonged expiration
- Intercostal and supraclavicular retractions
- Rhonchi

DX: Allergy skin testing, PFT, labs (CBC, ABG), CXR

TX: Avoidance of allergens, tobacco, and beta-adrenergic blockers; medication (inhaled beta₂ agonists, inhaled corticosteroids, leukotriene receptor agonists, systemic steroids [during infections and exacerbations]), peak expiratory flow monitoring

F/U: Reevaluation in 24 hours, then every 3 to 5 days, then every 1 to 3 months

ANAPHYLAXIS
Additional signs and symptoms
- Stridor
- Weakness
- Angioedema
- Intercostal retractions
- Nasal edema and congestion
- Watery rhinorrhea

DX: PE, history of allergen exposure

TX: Symptomatic treatment, airway and oxygenation maintenance, allergy testing (after treatment), medication (I.V. or Sub-Q epinephrine, antihistamines, nebulized albuterol)

F/U: Reevaluation within 24 hours

CHRONIC BRONCHITIS
Signs and symptoms
- Wheezing that varies in severity, location, and intensity
- Prolonged expiration
- Coarse crackles
- Scattered rhonchi
- Hacking, productive cough
- Dyspnea
- Clubbing
- Accessory muscle use
- Cyanosis
- Edema

DX: PFT, labs (CBC, ABG), CXR

TX: Smoking cessation, medication (pneumococcal vaccination, influenza vaccination, beta-₂ agonist therapy, bronchodilator, corticosteroids), avoidance of environmental irritants, avoidance of beta-adrenergic blockers and antihistamines, early treatment of infections, oxygen therapy

F/U: Return visit within 48 hours after acute exacerbation, then every 3 months

GERD
Signs and symptoms
- Hematemesis
- Abdominal pain
- Pyrosis
- Flatulence
- Dyspepsia
- Postural regurgitation

DX: Labs (electrolytes, CBC, stool guaiac), imaging studies (barium swallow, upper GI series, endoscopy), biopsy

TX: Diet and lifestyle modification, medication (histamine-2 blockers, antacids, proton pump inhibitors), blood transfusion (if indicated)

F/U: Reevaluation every 6 months unless the condition worsens, then referral to gastroenterologist

---

Additional differential diagnoses: aspiration of a foreign body ▪ aspiration pneumonitis ▪ bronchial adenoma ▪ bronchiectasis ▪ bronchogenic carcinoma ▪ chemical pneumonitis (acute) ▪ emphysema ▪ inhalation injury ▪ pneumothorax (tension) ▪ pulmonary coccidioidomycosis ▪ pulmonary edema ▪ pulmonary embolus ▪ pulmonary tuberculosis ▪ thyroid goiter ▪ tracheobronchitis ▪ Wegener's granulomatosis

English-Spanish phrases used in cardiac and respiratory assessment

Health problems

CHEST PAIN
General
Have you ever had chest pain or discomfort?

¿Ha tenido Ud. alguna vez dolor de pecho o molestia?

Description
  –How is the pain?
    Constant?
    Intermittent?
Do any activities produce the pain?

  –¿Cómo lo describiría?
    ¿Constante?
    ¿Intermitente?
¿Qué actividad o actividades producen el dolor?

Does it hurt when you breathe normally or when you breathe deeply?

¿Tiene Ud. el dolor cuando respira normalmente o cuando respira profundamente?

What does the pain feel like?
  –Crushing or squeezing?
  –Someone or something heavy pressing on your chest?
  –Pressure or tightness?
  –Dull ache?
  –Burning sensation?
  –Sharp or stabbing like a knife?
  – Ripping or tearing sensation?
  – Heartburn-type pain?

¿Qué clase de dolor?
  –¿Siente que lo aplasta u oprime?
  –¿Como si algo pesado estuviera oprimiendo su pecho?
  –¿Siente presión o tensión?
  –¿Es un dolor sordo?
  –¿Sensación de ardor?
  –¿Agudo o punzante como un cuchillo?
  –¿Siente que es un dolor desgarrante?
  –¿Siente como si le quemara el pecho?

Other symptoms
Do you have any other symptoms with the pain?
  –Anxiety?
  –Dizziness?
  –Fainting?
  –Feeling of doom?

¿Tiene otros sintomas con el dolor?

  –¿Ansiedad?
  –¿Mareos?
  –¿Desmayos?
  –¿Depresión?

–Jaw pain?   –¿Dolor de mandíbula?
–Nausea?   –¿Náuseas?
–Restlessness?   –¿Inquietud?
–Shortness of breath?   –¿Respiración entrecortada?
–Sweating?   –¿Sudor?
–Vomiting?   –¿Vómitos?
–Weakness?   –¿Debilidad?

Aggravating factors
Does anything make the pain worse?   ¿Hay algo que empeora el dolor?
–Standing?   –¿Pararse?
–Walking?   –¿Caminar?
–Climbing stairs?   –¿Subir escaleras?

Alleviating factors
Does anything make the pain better?   ¿Hay algo que alivia el dolor?
–Lying down?   –¿Acostarse?
–Resting?   –¿Descansar?
–Taking medication?   –¿Tomar medicamentos?

Location
Where in your chest do you feel the pain?   ¿En qué parte del pecho siente Ud. el dolor?

Can you point to where you feel the pain?   ¿Puede Ud. señalar con el dedo dónde siente el dolor?

–Does it radiate to any other area?   –¿Se irradia este dolor a otra parte del cuerpo?

Duration
How long have you been having this chest pain?   ¿Hace cuánto tiempo que Ud. tiene este dolor de pecho?
–Did it start recently?   –¿Comenzó hace poco?
–Over the last few hours, days, or weeks?   –¿Hace unas horas, días o semanas?
How long does an attack last?   ¿Cuánto tiempo dura el ataque?
Seconds?   –¿Segundos?
–Minutes?   –¿Minutos?
–Hours?   –¿Horas?
–Days?   –¿Días?

Arrhythmias
Does your heart ever feel like it's: ¿Siente Ud. alguna vez que el corazón:
–pounding?   –¿le late violentamente?
–racing?   –¿le late aceleradamente?
–skipping beats?   –¿se saltea latidos?

When does this feeling occur?	¿Cuándo ocurre esta sensación?
–While resting?	–¿Mientras descansa?
–During an activity?	–¿Mientras hace alguna actividad?
–After an activity?	–¿Después de desempeñar una actividad?
–After exercising?	–¿Después de hacer ejercicio?
–After walking up steps?	–¿Después de subir escalones?
–After eating?	–¿Después de comer?

COUGH
General

Do you have a cough?	¿Tiene Ud. tos?

Description

What does it sound like?	¿Qué sonido tiene?
Hacking?	¿Tos seca?
Barking?	¿Tos perruna?
Congested?	¿Congestionada?

Timing

Does it usually occur at a certain time of day?	Por lo general, ¿ocurre a cierta hora del día?
–When?	–¿Cuándo?

Sputum production

Do you cough up sputum?	¿Expectora Ud.?
–How much do you cough up each day?	¿Cuánto expectora Ud. al día?
–What color is it?	–¿De qué color es?
Red?	¿Rojo?
Pink?	¿Rosado?
With streaks of blood?	¿Sanguinolento?
Yellow?	¿Amarillo?
Green?	¿Verde?
White?	¿Blanco?
Clear?	¿Transparente o claro?
–How does it smell?	–¿Qué olor tiene?
–Is it thick?	–¿Es el esputo denso?
–Is it thin?	–¿Es el esputo claro?
–What time of day do you cough up the most sputum?	–¿A qué hora del día expectora Ud. más?
Morning?	¿Por la mañana?
Night?	¿Por la noche?
After meals?	¿Después de las comidas?

BREATHING DIFFICULTIES
General
Do you have shortness of breath?

¿Sufre Ud. de falta de aire?

Description
When did it occur?

¿Cuándo ocurrió?

Is it constant?

¿Es constante?

Is it intermittent?

¿Es intermitente?

What makes it better?

¿Qué lo mejora?

What makes it worse?

¿Qué lo empeora?

Does position, medication, or relaxation relieve it?

¿Se alivia Ud. con el descanso, algún medicamento, o cambio de postura?

Is shortness of breath related to any activity?

¿Está relacionado con alguna actividad?

–Which activity?

–¿Con qué actividad?

Is shortness of breath accompanied by coughing?

¿Va acompañado de tos?

Do your lips or nail beds ever turn blue?

¿Alguna vez se le ponen azules los labios o el lecho de la uña?

Does body position affect your breathing?

¿Le afecta la respiración la postura del cuerpo?

–How?

–¿Cómo?

Does time of day affect your breathing?

¿Le afecta la respiración la hora del día?

–What time of day?

–¿A qué hora del día se siente peor?

FATIGUE
General
Do you tire more easily than you used to?

¿Se cansa Ud. con más facilidad que antes?

What type of activity causes you to feel fatigued?

¿Qué tipo de actividad le hace sentirse fatigado(a)?

–How long can you perform this activity before you feel fatigued?

–¿Por cuánto tiempo puede Ud. hacer esa actividad antes de sentirse fatigado?

–Does rest relieve the fatigue?

–¿El reposo le mitiga la fatiga?

How many stairs can you climb before you feel short of breath?

¿Cuántos escalones puede Ud. subir antes de sentir falta de aire?

How many blocks can you walk before you feel short of breath?

¿Cuántas calles puede caminar antes de sentir falta de aire?

SWELLING
General
Do your shoes or rings feel tight?

¿Le aprietan los zapatos o los anillos?

Do your ankles or feet feel swollen?

¿Siente Ud. que se le hinchan los tobillos o los pies?

How long have you felt this way?

¿Hace cuánto tiempo que se siente así?

Have you gained or lost weight recently?	¿Aumentó o perdió peso recientemente?

RESPIRATORY TREATMENTS
Medications

Do you ever use over-the-counter nasal sprays or inhalers?	¿Usa Ud. alguna vez rociadores nasales o inhaladores?
–What kind do you use?	–¿Qué clase usa Ud.?
–How frequently do you use them?	–¿Con qué frecuencia lo usa Ud.?
Do you take any over-the-counter or prescription drugs for your respiratory difficulties?	¿Toma Ud. medicamentos con receta o sin receta médica para sus dificultades respiratorias?
–Which drugs?	–¿Qué medicamentos?
–Any steroids?	–¿Esteroides?
–How often do you take them?	–¿Con qué frecuencia los toma Ud.?
–When did you last take these drugs?	–¿Cuándo fue la última vez que tomó estos medicamentos?
Do you use a nebulizer or other breathing treatment?	¿Usa Ud. un nebulizador u otro tratamiento para respirar?
–What condition does it treat?	–¿Para qué enfermedad usa Ud. el tratamiento?
–What dose do you use?	–¿Qué dosis se le dió?
–How often do you have a treatment?	–¿Con qué frecuencia sigue Ud. un tratamiento?

Oxygen

Do you use oxygen at home?	¿Usa Ud. oxígeno en casa?
–Do you use a cannula or a mask?	–¿Usa Ud. una cánula o máscara?
–How often do you use it?	–¿Con qué frecuencia la usa?
Continuously?	¿Continuamente?
Intermittently?	¿Intermitentemente?

Medical history

CARDIOVASCULAR HISTORY
General

Were you born with a heart problem?	¿Nació Ud. con algún problema cardiaco?
–When was it treated?	–¿Cuándo recibió tratamiento?
–How was it treated?	–¿Cómo se le trató?

Rheumatic fever

Have you had rheumatic fever?	¿Ha tenido Ud. fiebre reumática?
When?	–¿Cuándo?
Have any heart problems resulted from the rheumatic fever?	¿Le han resultado enfermedades del corazón a causa de la fiebre reumática?

Heart murmur

Have you ever been told you had a heart murmur?
- –Who told you about it?
- –When did you find out about it?

¿Alguna vez se le ha dicho que tenía un soplo cardiaco?
- –¿Quién se lo dijo?
- –¿Cuándo se enteró Ud. de esto?

Other conditions

Do you have any of the following conditions?
- –High blood pressure?
- –High cholesterol?
- –Diabetes mellitus?
- –Heart failure?

When was the disorder first diagnosed?

How do you manage it?
How has it affected your lifestyle?

¿Tiene Ud. alguna de las siguientes enfermedades?
- –¿Presión sanguínea alta?
- ¿Colesterol alto?
- –¿Diabetes melitus?
- –¿Problemas cardiacos?

¿Cuándo se le diagnosticó por primera vez este trastorno?

¿Cómo se lo controla?
¿Cómo ha afectado su modo de vida?

Procedures

Have you ever had _____?

- –a cardiac catheterization
- –a stent placed
- –an angioplasty
- –any type of heart surgery
- –a pacemaker inserted
- –a defibrillator inserted

¿Alguna vez fue sometido(a) a _____?

- –una cateterización cardiaca
- –colocación de stent
- –una angioplastía
- –un tipo de cirugía cardiaca
- –colocación de marcapasos
- –colocación de desfibrilador

RESPIRATORY HISTORY
General

Have you had any lung problems?

- –Asthma?
- –Tuberculosis?
- –Pneumonia?
- –Influenza?
- –Sinus problems?
- –Emphysema?
- –Allergies?
- –Other?

How long did the problem last?
How was the problem treated?

¿Ha tenido Ud. problemas de los pulmones?

- –¿Asma?
- –¿Tuberculosis?
- –¿Pulmonía?
- –¿Gripe?
- –¿Sinusitis?
- –¿Enfisema?
- –¿Alergias?
- –¿Otros?

¿Cuánto tiempo le duró el problema?
¿Qué tratamiento recibió?

Exposure history

Have you been exposed to anyone with a respiratory disease?

¿Ha estado Ud. expuesto(a) a alguna persona que tenga una enfermedad respiratoria?

–What type of disease?

–¿Qué clase de enfermedad?

–When were you exposed?

–¿Cuándo estuvo Ud. expuesto(a)?

Are there any respiratory irritants in your home, such as fresh paint, cleaning sprays, or heavy cigarette smoke?

¿Hay en su casa agentes irritantes que le afectan la respiración, tales como pintura fresca, nebulización de productos de limpieza o humo de cigarrillos?

Diagnostic tests

Have you had a diagnostic study of the lungs?

¿Le han hecho una revisión de los pulmones?

–What type?

–¿De qué tipo?

–Why did you have it?

–¿Por qué tuvo la cirugía o por qué se le hizo el estudio?

When was your last chest X-ray?

¿Cuándo se le tomó la última radiografía de los pulmones?

When was your last tuberculosis test?

¿Cuándo se le hizo el último análisis para la tuberculosis?

–What was the result?

–¿Cuál fue el resultado?

Procedures

Have you had chest surgery?

¿Lo han operado del pecho?

Have you had surgery on your___?

¿Fue operado_____?

–lungs

–de los pulmones

–sinuses

–de los senos nasales

–mouth, nose, or throat

–de la boca, nariz o garganta

Allergies

Do you have allergies that flare up in different seasons?

¿Tiene Ud. alergias que se agravan durante diferentes temporadas del año?

–What causes them?

–¿Qué es lo que las causa?

–Do they cause any of these symptoms:

–¿Le causan alguno de los siguientes síntomas:

runny nose?

¿le gotea la nariz?

itching eyes?

¿picazón en los ojos?

congestion?

¿congestión?

other symptoms?

¿otros síntomas?

–What do you do to relieve these symptoms?

–¿Qué hace Ud. para aliviar estos síntomas?

In the last 2 months, have you had:

¿En los últimos dos meses ha tenido:

–fever?

–¿fiebre?

–chills?

–¿escalofríos?

–fatigue?

–¿fatiga?

–night sweats?

–¿sudores nocturnos?

Family history

Has anyone in your family been treated for heart disease?	¿Algún miembro de su familia ha recibido tratamiento para alguna enfermedad cardiaca?
Has anyone in your family died suddenly of an unknown cause?	¿Algún miembro de su familia ha muerto repentinamente por causa desconocida?
Has any member of your family had:	¿Algún miembro de su familia tuvo alguna de las siguientes enfermedades:

-emphysema?
-asthma?
-respiratory allergies?
-tuberculosis?
-sarcoidosis?

Did you have contact with the family member who had tuberculosis?
-When?
-Do they live with you?

—¿enfisema?
—¿asma?
—¿alergias del sistema respiratorio?
—¿tuberculosis?
—¿sarcoidosis?

¿Estuvo Ud. en contacto con el miembro de la familia que tuvo tuberculosis?
—¿Cuándo?
—¿Vive con Ud.?

Lifestyle

TOBACCO

Do you smoke or chew tobacco?
-What do you smoke?
  Cigarettes?
  Cigars?
  Pipes?
-How long have you smoked or chewed tobacco?
-How many cigarettes, cigars, or pipes of tobacco do you smoke per day?

¿Ud. fuma o masca tabaco?
  —¿Qué fuma?
    ¿Cigarrillos?
    ¿Cigarros (puros)?
    ¿Pipas?
  —¿Hace cuánto tiempo que fuma o masca tabaco?
  —¿Cuántos cigarrillos, cigarros (puros) o pipas de tabaco fuma Ud. al día?

ALCOHOL

Do you drink alcoholic beverages?
-What type?
  Beer?
  Wine?
  Hard liquor?
How often do you drink?
How many drinks per day?

¿Toma Ud. bebidas alcohólicas?
  —¿Qué clase?
    ¿Cerveza?
    ¿Vino?
    ¿Aguardiente?
¿Con qué frecuencia bebe Ud.?
¿Cuántos tragos bebe al día?

SLEEP PATTERNS

Do you awaken during the night to urinate?

¿Se despierta Ud. durante la noche para orinar?

Do you ever have shortness of breath or coughing spells during the night?

¿Alguna vez tiene la respiración entrecortada o accesos de tos por la noche?

Do you become short of breath when you lie flat?

¿Le falta a Ud. la respiración cuando se acuesta de espaldas?

How many pillows do you use at night?

¿Cuántas almohadas usa Ud. en la noche?

Have your sleep patterns changed because of breathing problems?

¿Han cambiado sus hábitos de dormir a causa de sus problemas respiratorios?

EXERCISE

Do you exercise routinely?

¿Hace Ud. ejercicio rutinariamente?

–How often do you exercise?

–¿Con qué frecuencia hace ejercicio?

–How intensely do you exercise?

–¿Con qué intensidad hace el ejercicio?

–How long do you spend exercising?

–¿Por cuánto tiempo hace Ud. ejercicio?

STRESS

How often do you experience stress?

¿Con qué frecuencia siente Ud. estrés?

–Rarely?

–¿Raramente?

–Once in awhile?

–¿De vez en cuando?

–Very often?

–¿Con frecuencia?

Are you ever overwhelmed by stress?

¿Se siente Ud. alguna vez abrumado(a) por el estrés?

EMPLOYMENT

Are you currently employed?

¿Tiene Ud. empleo actualmente?

What's your occupation?

¿Cuál es su profesión o trabajo?

What were your previous occupations?

¿Qué otros empleos ha tenido anteriormente?

Are you exposed to any known respiratory irritants at work?

¿Sabe Ud. si está expuesto en su trabajo a agentes irritantes que afecten su respiración?

–Do you use safety measures during exposure?

–¿Usa Ud. medidas de seguridad mientras está expuesto(a)?

Posttest answers

Heart sounds

1 THE HEART AND AUSCULTATION

1. The heart's primary function is to pump deoxygenated blood to the lungs, where carbon dioxide–oxygen (CO_2-O_2) exchange takes place, and to pump oxygenated blood throughout the body.

2. The top, or superior, part of the heart is called the base; the most inferior and lateral part, the apex.

3. Anatomically, the heart is divided into two functional sides: the right side and the left side, which are separated by the septum. Each side is further divided into chambers. The right atrium and right ventricle are the right-sided chambers; the left atrium and left ventricle, the left-sided chambers.

4. The right atrioventricular, or tricuspid, valve separates the right atrium and right ventricle, and the left atrioventricular, or mitral, valve separates the left atrium and left ventricle. The semilunar pulmonic valve separates the right ventricle and pulmonary artery, and the semilunar aortic valve separates the left ventricle and aorta.

5. Deoxygenated, or venous, blood from nearly every tissue of the body flows into the right atrium via the superior and inferior venae cavae. Venous blood from the heart tissues itself drains via the coronary sinus directly into the right atrium. The deoxygenated blood collects there until the tricuspid valve opens. Then the deoxygenated blood flows into the right ventricle. From the right ventricle, it flows through the open pulmonic valve into the pulmonary artery, which distributes deoxygenated blood to both lungs.

   The complex process of CO_2-O_2 exchange occurs in the alveoli. Then the oxygenated blood flows back to the heart via the pulmonary veins and collects in the left atrium. From there, oxygenated blood flows through the opened mitral valve into the left ventricle. When the aortic valve opens, oxygenated blood flows into the aorta, which transports it to all body tissues.

6. A healthy heart in a 154-lb (70-kg) adult pumps about 1⅝ gal (6 L) of blood per minute. Because the total blood volume in a healthy adult is about 1⅜ gal (5.2 L), the heart pumps the body's total blood volume in less than 1 minute.

7. During systole, heart muscle contraction raises the intracardiac pressures enough to systematically open the valves and pump a certain volume of blood. During diastole, the heart muscle relaxes, allowing the chambers to fill with blood again. This entire sequence is called the cardiac cycle.

8. Cardiac output, the amount of blood pumped into the aorta each minute, depends largely on stroke volume. Stroke volume is the amount of blood pumped during each ventricular contraction.

9. Electrical impulses are primarily responsible for the heart's rhythmic pumping action. These impulses are discharged automatically from specialized cardiac cells that stimulate the heart muscle to contract. The specialized cardiac cells and fibers within the conduction system work together to produce a contraction. The impulses fired from the sinoatrial (SA) node, located just below the entrance of the superior vena cava in the right atrium, normally initiate the cardiac cycle, and they usually travel along specific pathways. For example, when the electrical impulse leaves the SA node, it travels along the atrial conduction pathways located in the atrial walls, initiating atrial systole (or atrial contraction). The impulse rapidly reaches the atrioventricular (AV) node, located near the AV junction. From the AV node, the impulse travels to the bundle of His, located in the interventricular septum. Here the bundle of His divides into the right and left bundle branches. These branches further divide into the tiny Purkinje fibers located throughout the ventricular walls. As the impulse travels through the bundle branches and Purkinje fibers, it initiates ventricular systole (or ventricular contraction).

10. The sequential and synchronized contraction (systole) and relaxation (diastole) of the atria and ventricles,

along with the opening and closing of competent, properly functioning valves, ensure a unidirectional flow of blood through the heart and allow the heart to effectively move blood to all areas of the body quickly and repeatedly.

11. The aortic area is located at the second right intercostal space close to the sternal border. Abbreviated as 2RSB, this area is where sounds of the aortic valve and aorta are heard best.

    The pulmonic area is located at the second left intercostal space close to the sternal border. Abbreviated as 2LSB, this area is where sounds of the pulmonic valve and pulmonary artery are heard best.

    The tricuspid area is located at the fifth left intercostal space close to the sternal border. This area, abbreviated as 5LSB, is where sounds of the tricuspid valve and right ventricle are heard best.

    The mitral area is located at or near the fifth intercostal space just medial to the left midclavicular line. This area, directly over the left ventricle, is sometimes referred to as the apical area, or apex. The mitral area, abbreviated as 5LMCL, is where sounds of the mitral valve and left ventricle are heard best.

    Erb's point differs from the four other areas in that it isn't named after a heart valve, but it's a good location to hear sounds of aortic and pulmonic origin. Abbreviated as 3LSB, this area is located at the third left intercostal space close to the sternal border.

12. The chest piece should have a diaphragm and a bell. The diaphragm is designed to transmit high-pitched sounds more clearly, and the bell is designed to transmit low-pitched sounds more clearly. A flat, adult-sized diaphragm should be about 1⅜"(3.5 cm) in diameter and should be smooth, thin, and stiff enough to filter out low-frequency sounds. The bell should be about 1" (2.5 cm) in diameter and deep enough so that it doesn't fill with tissue when placed on the chest wall. The tubing length can vary from 10" to 15" (25 to 38 cm), and its inside diameter should be ⅛" — or ³⁄₁₆" if the tubing is longer. The earpieces should fit inside the ears comfortably, and the stethoscope shouldn't be too heavy because its weight can interfere with your ability to hear sounds accurately.

13. To use the diaphragm, firmly grasp the metal area between the bell and the diaphragm with your finger and

thumb, and press down firmly on the chest wall. Or place your fingertips on the bell's rim, and press down firmly against the chest wall. Apply enough pressure so that an indentation remains on the skin after you remove the stethoscope.

The bell, however, should touch the chest wall only lightly. If you exert too much pressure when listening with the bell, the stretched skin beneath it will act as a diaphragm, filtering out low-pitched sounds. To hold the bell correctly, grasp the diaphragm's outer edges with the index finger and thumb, and gently rest the bell on the chest wall. Look at the skin around the edges of the bell to see if too much pressure is being applied. If any signs of indentation appear on the skin, relax the downward pressure.

14. Listen first with the diaphragm over the aortic area. Try to recognize the characteristic *lub-dub* sound of each cardiac cycle. After listening through several cycles, inch the diaphragm toward the pulmonic area. Try to move the diaphragm without losing track of the *lub-dub*. Then listen over the pulmonic area, Erb's point, and the tricuspid area. Continue in this manner until you reach the mitral area.

15. To locate the point of maximum impulse (PMI), turn the patient slightly to the left, placing him in the partial left lateral recumbent position. This position brings the heart's apex closer to the chest wall. Closely observe the chest wall for any sign of the heartbeat, which may appear as a rhythmic bulging. Then use your fingertip pads to palpate for the heartbeat between the fourth and the sixth intercostal spaces, near the left midclavicular line. The chest wall site where the heartbeat is seen and felt is the PMI.

16. Several maneuvers can enhance the heart sounds heard during auscultation. For example, you can have the patient squat or stand, hold his breath, or raise his legs while in a supine position. Another common technique is to have the patient cough several times or perform Valsalva's maneuver. During this maneuver, the patient attempts to exhale forcefully against a closed glottis.

Another method that enhances the sounds heard during auscultation is to have the patient increase the depth of breathing. This decreases the respiratory rate and makes it easier to distinguish sounds originating in either the right or left side of the heart. For example, during inspiration, venous return is enhanced, accentuating right-sided cardiac events; during expiration, left-sided blood flow is enhanced, accentuating left-sided cardiac events.

You can help the patient by raising your hand slowly when you want him to inhale, and then lowering your hand slowly when you want him to exhale. Caution the patient to breathe smoothly and continuously. Breath holding may produce a Valsalva-type maneuver, which will negate the cardiac responses being assessed.

2 HEART SOUND DYNAMICS

1. The heart sounds heard through a stethoscope during auscultation are generated by vibrations from the heart's walls and valves.

2. At the beginning of ventricular diastole, the ventricles are relaxed, and the aortic valve closes slightly before the pulmonic valve. For a brief time, both the atrioventricular and semilunar valves are closed. During this period, called isovolumic relaxation, intraventricular pressures continue to fall. But the opening of the tricuspid and mitral valves follows quickly, and blood begins to flow passively from the atria into the ventricles. As the ventricles become distended, the rate of ventricular filling decreases.

3. When the atria contract (from stimulation of the atrial myocardium by an electrical impulse from the sinoatrial node), a slight boost in ventricular volume occurs. The resulting increase in ventricular pressures causes the mitral and tricuspid valves to begin closing. During atrial systole, the electrical impulse moves more slowly through the atrioventricular node and into the bundle of His. The impulse proceeds rapidly through the bundle branches and the Purkinje fibers, activating ventricular systole.

4. Early in ventricular systole, intraventricular pressures increase above intra-atrial pressures, and the mitral and tricuspid valves close suddenly. Normally, the mitral valve closes slightly before the tricuspid valve. For another brief period, both the atrioventricular and semilunar valves are closed. During this period, known as isovolumic contraction, ventricular pressures increase, becoming higher than the pressures in the pulmonary artery and aorta. The aortic and pulmonic

valves open, and blood is ejected through them. Normally, the aortic valve opens slightly before the pulmonic valve.

After a period of maximum ejection of blood from the ventricles, the ventricles begin to relax. As intraventricular pressures fall rapidly, a brief backflow of blood from the pulmonary artery and aorta toward the ventricles occurs. Reduced intraventricular pressures and the temporary backflow of blood are associated with closure of the aortic and pulmonic valves at the end of systole.

5. S_1, the first heart sound, is heard at the beginning of systole. It's associated with closure of the mitral and tricuspid valves and the increasing pressures within the ventricles that cause the moving valve leaflets and cord structures to decelerate. S_1 is generated by the closing atrioventricular valves and the vibrations associated with tensing of the chordae tendineae and ventricular walls.

6. At the end of ventricular systole, ventricular pressures fall rapidly, causing a slight backflow of blood from the aorta and pulmonary artery. The decrease in ventricular pressures to a level below that of aortic and pulmonic systolic pressures and the temporary backflow of blood and recoil events are associated with closure of the aortic and pulmonic valves. The vibrations associated with these events produce the second heart sound, S_2, which marks the end of ventricular systole.

7. A sound's location is the anatomic area on the patient's chest wall where the sound is heard best. Bony structures and landmarks, such as the sternum and left midclavicular line, are used to describe the precise location.

8. Intensity refers to the sound's loudness during auscultation. Usually, a sound's intensity is determined subjectively, based on experience. However, when abnormalities are present, intensity can be determined electronically by recording a phonocardiogram and measuring the amplitude of the sound's vibrations.

9. Sound duration refers to the length of time the sound is heard; it can be described as either short or long. A sound's duration affects whether you hear it as a click, a thud, or a snap.

10. A sound's pitch is determined by the frequency of its vibrations. High-frequency sounds, like the notes of a

piccolo, are best heard with the diaphragm of the
stethoscope; low-frequency sounds, like the notes of a
tuba, are best heard with the bell of the stethoscope.

11. Sound quality is determined by the combination of its
frequencies. It may be described as sharp, dull, booming, or snapping.

12. A sound's timing refers to when it is heard during the
cardiac cycle (during systole or diastole).

13. The sequential and rhythmic process of depolarization
and repolarization is referred to as the heart's electrical
activity. Electrical activity occurs at the cellular level
in every contractile cell of the myocardium. Initially,
the contractile cell is polarized (in a resting state). Depolarization begins when the cell membrane becomes
permeable to the flow of sodium ions into the cell. Repolarization begins when calcium ions move into the
cell and potassium ions begin to move out of the cell.
Then the accumulated intracellular sodium and calcium ions are extruded while the lost potassium is restored to the cell by the sodium-potassium pump, completing repolarization. The cell is polarized to its original ionic state, and the myocardium relaxes.

14. The electrocardiogram (ECG) correlates with the
heart's electrical activity in the following manner: The
sinoatrial node fires, and the impulse spreads through
the atria. The P wave, part of the ECG waveform, represents atrial depolarization. Atrial contraction is stimulated by—and closely follows—atrial depolarization. The QRS complex represents electrical depolarization of the ventricles. Atrial repolarization isn't seen
in the ECG because it's hidden in the PR segment and
the QRS complex. The T wave represents ventricular
repolarization. The impulse travels through the atrioventricular node, the bundle of His, the bundle
branches, and the Purkinje fibers before the ventricles
contract. The time between atrial depolarization and
ventricular depolarization is recorded in the ECG as
the PR interval: It begins at the onset of the P wave
and lasts until the onset of the QRS complex. The time
interval of the PR interval correlates with the time interval between atrial contraction and ventricular contraction.

15. The first (S_1) and second (S_2) heart sounds directly
correlate with a patient's ECG. S_1 normally occurs just

after the QRS complex; S_2 occurs at the end of the T wave.

16. The intensity, duration, and timing of heart sounds can be represented graphically as rectangular blocks placed perpendicularly on a horizontal baseline. The block's height corresponds to the heart sound's intensity (the higher the intensity, the higher the block); the block's width corresponds to the sound's duration. The proximity of one block on the baseline to the next block represents the time interval between one sound and another. The spacing between the blocks can also be used to specify the relationship of a particular heart sound to systole or diastole.

3 THE FIRST HEART SOUND

1. The cardiac vibrations associated with closure of the mitral and tricuspid valves produce the first heart sound, S_1.

2. Normally, only two components of the first heart sound are audible. The first component, referred to as M_1, is associated with closure of the mitral valve; the second component, T_1, is associated with closure of the tricuspid valve. Both valves close at the beginning of ventricular systole, but the mitral valve usually closes slightly ahead of the tricuspid valve.

3. M_1 and T_1 are usually perceived as a single sound called S_1, which is heard best near the heart's apex over the mitral area using the diaphragm of the stethoscope. A single sound is heard because left heart sounds are normally more intense at this site. S_1 occurs shortly after the beginning of the QRS complex in the ECG waveform. As you inch the stethoscope from the mitral area toward the tricuspid area, without losing track of S_1, the M_1 and T_1 components of S_1 become evident. Expiration may make them easier to hear. T_1 trails M_1 slightly and is softer; it's heard best near the left sternal border. The timing of M_1 and T_1 with the QRS complex remains the same. In many patients, only one first heart sound is heard because M_1 and T_1 are separated by ≤ 20 milliseconds, which the human ear perceives as one sound.

4. S_1 is usually heard best near the heart's apex over the mitral area. Its intensity directly relates to the force of ventricular contraction and the ECG PR interval. The shorter the PR interval, the more widely open are the

mitral and tricuspid leaflets at the onset of ventricular contraction. Thus the distance they must be moved is greater, resulting in more intense vibrations when they close and a louder S_1. S_1 is short in duration. You can hear its high pitch best using the diaphragm of the stethoscope. The quality of S_1 is somewhat dull. S_1 timing coincides with the beginning of ventricular systole and just precedes a palpable carotid pulse.

5. S_1 can be enhanced by sympathetic stimulation, such as that provided by a brief period of exercise.

6. The normal M_1-T_1 split heard over the tricuspid area widens when electrical activation and contraction of the right ventricle are delayed. Such a delay causes delayed tricuspid valve closure.

7. Conditions associated with an abnormal S_1 split are complete right bundle-branch block, left ventricular ectopic beats, epicardial pacing of the left ventricle, tricuspid stenosis, atrial septal defect, Ebstein's anomaly, and left ventricular tachycardia.

4 THE SECOND HEART SOUND

1. The cardiac vibrations associated with closure of the aortic and pulmonic valves produce the second heart sound, S_2.

2. S_2, like S_1, has two basic components: the aortic (A_2) component and the pulmonic (P_2) component. Both valves close at the end of ventricular systole. Normally, the aortic valve closes slightly ahead of the pulmonic valve.

3. To hear both components of S_2, listen carefully over the pulmonic area using the diaphragm of the stethoscope. Listen for S_2 just after the T wave in the patient's ECG.

4. S_2 is usually heard best near the heart's base over the pulmonic area or over Erb's point. Its intensity directly relates to the amount of closing pressure in the aorta and pulmonary artery. S_2 is slightly shorter in duration than S_1. You can hear its high pitch best using the diaphragm of the stethoscope. The quality of S_2 is somewhat booming; its timing coincides with the end of ventricular systole.

5. The splitting of S_2 into the A_2 and P_2 components is heard best during inspiration over Erb's point. Remember, inspiration causes an increase in venous return to the right side of the heart. This increased venous re-

turn prolongs right ventricular ejection time. Also, inspiration reduces pressure in the pulmonary artery. Both result in a delayed P_2, or closure of the pulmonic valve. A decrease in blood flow to the left side of the heart occurs simultaneously, resulting in a shorter left ventricular ejection time. Therefore, A_2 is heard earlier than P_2. This means that a normal S_2 split is heard during inspiration, and the A_2 and P_2 components fuse during expiration.

6. The intensity of A_2 and P_2 changes proportionally with the difference in pressure gradients across the closed aortic and pulmonic valves. For example, P_2 may be louder than normal in conditions associated with elevated pulmonary artery diastolic pressure, as occurs in some patients with heart failure, mitral stenosis, Eisenmenger's syndrome, or other congenital heart diseases. When P_2 increases in intensity, it's sometimes heard over the mitral area and along the left sternal border.

   A_2 intensity increases when diastolic pressure in the aorta increases. This is common during exercise; during states of excitement, such as extreme fear; in hyperkinetic conditions, such as thyrotoxicosis, fever, and pregnancy; and in systemic hypertension. However, if the patient has left ventricular decompensation, ventricular relaxation is slower, and the pressure gradients may not be great enough to produce an accentuated A_2. A_2 is also softer or absent when aortic valve motion is restricted, as in severe aortic stenosis.

   Conversely, A_2 intensity may be diminished in conditions that alter the development of diastolic pressure gradients, such as aortic regurgitation and hypotension. A_2 intensity also decreases when ventricular dysfunction is present, such as after an acute myocardial infarction. In this condition, P_2 may become louder if pulmonary artery pressure rises and systemic pressure falls.

7. Abnormal splitting is related to valvular dysfunction, to alterations in blood flow to or from the ventricles, or to both. These changes may cause the normal S_2 split to be absent during both phases of the respiratory cycle. Thus, only a single S_2 is heard over Erb's point. In another case, the split sounds may persist through inspiration and expiration with little or no respiratory variation. The split sounds also may be heard paradox-

ically on expiration. The A_2-P_2 intervals vary, as does the intensity of A_2 and P_2 during the respiratory cycle. Changes in S_2 splits are usually most noticeable at the beginning of inspiration and expiration.

8. The P_2 component may not be heard during auscultation over Erb's point in the patient with severe pulmonic stenosis. Consequently, S_2 remains a single sound during both inspiration and expiration. A normal S_2 split may also be absent if the A_2 sound masks the P_2 sound or vice versa — for example, when one sound is significantly louder than the other, making splitting inaudible. This phenomenon occurs in patients with pulmonary hypertension. On the other hand, systemic hypertension causes A_2 to be delayed and to fuse with P_2 during inspiration.

9. A persistent S_2 split occurs when A_2 and P_2 don't fuse into one sound during expiration, even though some respiratory variation in the intensity of A_2 and P_2 is heard. This persistent A_2-P_2 splitting during expiration usually results from early aortic valve closure or delayed pulmonic valve closure. Early aortic valve closure is associated with shortened left ventricular systole, which occurs in patients with mitral regurgitation, ventricular septal defects, or cardiac tamponade. Delayed pulmonic valve closure occurs when right ventricular systole is prolonged in patients with chronic pulmonary hypertension. In these patients, the A_2-P_2 split is heard, but the interval between the A_2 and P_2 components is narrowed.

    Another cause of persistent A_2-P_2 splitting throughout expiration is delayed electrical activation of the right ventricle, which delays P_2. This phenomenon is commonly found in patients with right bundle-branch block, left ventricular epicardial pacing, or left ventricular ectopic beats.

10. A prolonged right ventricular ejection time produces expiratory A_2-P_2 splits that are widened and that persist even when the patient is seated. This occurs in patients with atrial septal defects, acute pulmonary hypertension secondary to massive pulmonary emboli, or pulmonic stenosis. The P_2 component may not be audible at all in patients with severe pulmonic stenosis. A wide, fixed S_2 split is sometimes associated with the pulmonary vascular bed's increased capacitance (ability to receive blood volume and the decreased resis-

tance that accompanies it). This phenomenon occurs in patients with idiopathic dilation of the pulmonary artery or with atrial septal defects. This split doesn't change with respiration.

11. In a paradoxical, or reversed, S_2 split, P_2 precedes A_2, and the split sounds are heard during expiration instead of inspiration. This phenomenon is almost always caused by delayed aortic valve closure. If A_2 is delayed during expiration, it may follow P_2, causing an S_2 split; if A_2 is delayed during inspiration, A_2 and P_2 fuse because inspiration normally delays P_2. Therefore, S_2 is heard as a single sound during inspiration instead of a normally split sound.

   Delayed A_2 is commonly seen in patients with delayed activation of the left ventricle caused by left-bundle-branch block. It's also associated with right ventricular ectopic beats or right ventricular endocardial pacing. Prolonged left ventricular systole can also cause a delay in A_2. The same paradoxical P_2-A_2 split may be heard in patients with left-sided heart failure or ischemic heart disease.

   Left ventricular pressure overload, which occurs in patients with systemic hypertension or hypertrophic cardiomyopathy, may also cause paradoxical S_2 split. Likewise, aortic stenosis may lead to paradoxical splitting of A_2 and P_2; however, if the stenosis is severe, the A_2 component may not be audible. With left ventricular volume overload, which is commonly associated with aortic regurgitation or patent ductus arteriosus, A_2 may be delayed, and a paradoxical S_2 split may be heard.

5 THE THIRD AND FOURTH HEART SOUNDS

1. The two left ventricular diastolic filling sounds, S_3 and S_4, are sometimes heard over the mitral area. These sounds differ from S_1 and S_2 in that they're low-frequency sounds and produced by ventricular filling rather than associated with valve closure.

2. Early in diastole, after isovolumic relaxation, the mitral and tricuspid valves open and the ventricles fill and expand. In children and young adults, the left ventricle is normally compliant, permitting rapid filling. The left ventricle responds to this rapid filling with an abrupt change in wall motion that causes a sudden de-

crease in blood flow. These events generate vibrations that are responsible for the physiologic S_3.

3. A physiologic S_3 is commonly heard in patients with high-output conditions, in which rapid ventricular expansion, caused by increased blood volume, is present. Anemia, fever, pregnancy, and thyrotoxicosis are some of the conditions that cause rapid ventricular expansion, resulting in an S_3. An S_3 is also commonly heard in young, slender individuals during periods of excessive catecholamine release.

4. S_3 is usually heard best over the mitral area using the bell of the stethoscope, and it can typically be palpated over the same area. It's heard best during expiration, when blood flow into the left ventricle is increased.

5. S_3 is usually heard best near the heart's apex over the mitral area. In some patients, it's faint in intensity and difficult to hear; in others, it's loud and easy to hear. S_3 has a short duration, and it may occur only intermittently during every third or fourth heartbeat. It has a low pitch that's heard best using the bell of the stethoscope. S_3 usually has a dull, thudlike quality. Its timing is closely related to S_2, which is heard just after the T wave; S_3 follows S_2 by less than 0.2 second.

6. Because S_3 is associated with blood volume and velocity, it can be intensified by maneuvers that increase stroke volume, such as elevating the patient's legs from a recumbent position or having the patient exercise briefly or cough several times. Placing the patient in the partial left lateral recumbent position may enhance your ability to auscultate and palpate an S_3.

7. The differences between the two sounds are related to the patient's age, clinical condition, or both. Also, an S_3 gallop rhythm usually persists despite maneuvers that decrease venous return.

8. An abnormal S_4 may also be heard in conditions associated with increased blood volume and increased inflow velocity into the left ventricle. It's heard with or without an increase in ventricular diastolic pressure. Consequently, patients with mitral regurgitation and heart failure have an abnormal S_3. This heart sound can also be heard during increased blood flow through the mitral valve, which occurs in patients with ventricular septal defects, patent ductus arteriosus, or severe aortic regurgitation.

An abnormal S_3 in patients with constrictive pericarditis is called a pericardial knock. This type of abnormal S_3 occurs closer to S_2. The interval between S_2 and S_3 is usually less than 0.14 second.

9. An S_3 sometimes originates in the right ventricle instead of the left; when it does, it's always considered abnormal. A right-sided S_3 is heard best over the third, fourth, and fifth intercostal spaces along the left sternal border or over the epigastric area. It's more prominent during inspiration because of increased blood flow into the right ventricle.

10. By the end of diastole, the ventricles are nearly full; atrial contraction further stretches and fills the ventricles. The vibrations caused by this filling and stretching in late diastole generate an additional heart sound, S_4, sometimes called an atrial diastolic gallop.

11. S_4 is usually heard best near the heart's apex over the mitral area; occasionally, it's also palpable over this area. In some patients, it's faint in intensity and difficult to hear, but in others it's loud and easily heard. S_4 is relatively short in duration and may occur only intermittently, during every third or fourth heartbeat. It has a low pitch that's heard best using the bell of the stethoscope, and it has a thudlike quality. Its timing is presystolic: S_4 precedes S_1 and occurs during the PR interval.

12. To enhance S_4, place the patient in the partial left lateral recumbent position, which brings the heart closer to the chest wall.

13. An abnormal S_4 is almost always associated with increased mean left atrial pressure caused by a noncompliant left ventricle. This auscultatory finding is heard in patients with hypertension, hypertrophic cardiomyopathy, cardiomyopathies, or ischemic heart disease and during or after an acute myocardial infarction. When an S_4 is heard in a patient with hypertension, the systolic blood pressure usually exceeds 160 mm Hg or the diastolic pressure exceeds 100 mm Hg. An abnormal S_4 can also accompany volume overload conditions, such as hyperthyroidism, severe anemia, and sudden severe mitral regurgitation.

14. Normally, S_4 precedes S_1 by an appreciable interval that correlates with the PR interval on the ECG. However, in patients with first-degree AV block, the P wave occurs early in diastole, and S_4 may occur during the

early rapid diastolic filling period. Likewise, in tachycardia, S_4 may be superimposed on S_3 during early rapid filling. Should either condition exist, the S_4 fuses with S_3 to become a single diastolic filling sound called a summation gallop, which may be louder than S_4, S_3, or S_1.

15. An S_4 generated in the right ventricle is called a right-sided S_4. It's commonly heard in conditions that increase pressure in the right ventricle by more than 100 mm Hg, such as pulmonic stenosis or pulmonary hypertension. This heart sound is heard best over the third, fourth, and fifth intercostal spaces along the left sternal border and over the epigastric area with the patient in a supine position; it's more audible during inspiration.

16. Distinguishing an S_1 split (M_1-T_1) from an S_4-S_1 sequence is sometimes difficult; however, here are some ways to help you do so. An S_1 split is heard best between the mitral and tricuspid areas with the diaphragm of the stethoscope, whereas an S_4 is heard best over the mitral area using the bell of the stethoscope and usually isn't audible with the diaphragm. Also, an S_4 may be palpable. The intensity of S_4 can usually be increased by maneuvers that increase left atrial pressure, such as handgrip exercises.

17. S_3, like S_4, is low pitched and may be faint or loud and heard only intermittently. Both sounds are heard best over the mitral area using the bell of the stethoscope, with the patient in a partial left lateral recumbent position. Both sounds are intensified by expiration and can be enhanced by maneuvers that increase stroke volume, such as elevating the legs while in a recumbent position.

6 OTHER DIASTOLIC AND SYSTOLIC SOUNDS

1. If the mitral valve leaflets become stenotic or abnormally narrowed while remaining somewhat mobile, they create an opening snap (OS).

2. An OS is usually heard best near the heart's apex over the mitral area or just medial to it. Its intensity varies among patients, and it's usually easy to hear during auscultation. An OS has a short duration. It has a high pitch that's heard best using the diaphragm of the stethoscope, and it has a sharp, snaplike quality. Its

timing is closely related to S_2: An OS occurs early in ventricular diastole, just after the stenotic mitral valve opens.

3. One characteristic of an OS that helps distinguish it from P_2 is its timing: The A_2-P_2 interval is normally shorter than the A_2-OS interval. Also, when the patient stands, the A_2-P_2 interval narrows, whereas the A_2-OS interval widens. Another characteristic is that the A_2-OS interval remains constant throughout respiration, whereas the A_2-P_2 interval normally widens during inspiration and narrows during expiration. During inspiration, three distinct sounds can usually be heard over the pulmonic area; therefore, the sequence must be A_2, P_2, OS. In contrast, during expiration, the A_2-P_2 interval narrows or fuses, forming one sound. This creates an S_2-OS interval. Finally, P_2 usually isn't heard over the mitral area, so if you hear a split S_2 over this area, it is likely to be an S_2 and an OS.

4. One characteristic of an OS that helps distinguish it from an S_3 is its timing: The A_2-S_3 interval is usually longer than the A_2-OS interval. Also, S_3 is a low-frequency sound that's heard best over the mitral area using the bell of the stethoscope. In contrast, an OS produces a high-frequency sound that's more widely transmitted across the precordium and is heard best using the diaphragm of the stethoscope.

   Another characteristic of an OS is that its intensity usually isn't affected by having the patient stand, whereas S_3 intensity can be decreased by standing. If the OS is affected, it's intensified. Another way to differentiate S_3 from an OS is by listening for respiratory changes. An S_3 is usually louder during expiration than during inspiration and, in many patients, is palpable; an OS doesn't vary in intensity with respiration. Finally, if the murmur typically heard in patients with mitral stenosis is present, you can confirm that the sound is an OS.

5. A systolic ejection sound (ES) is usually a brief, high-frequency sound that's heard best using the diaphragm of the stethoscope. It may be heard near the heart's base over the aortic or pulmonic area, over Erb's point, or near the apex over the mitral area. A systolic ES occurs just after the QRS complex on the ECG waveform.

6. A pulmonic ejection sound (PES) is commonly associated with ventricular ejection and the maximum opening of a stenotic, yet mobile, pulmonic valve. It may also be caused by sudden distention of an already dilated pulmonary artery and by forceful ventricular ejection from pulmonary hypertension.

7. A PES is usually heard best near the heart's base over the pulmonic area. Its intensity is soft but may be equal to or greater than that of S_1. A PES has a short duration. It has a high pitch that's heard best using the diaphragm of the stethoscope, and it has a sharp, or clicklike, quality. Its timing is closely related to S_1: It occurs early in ventricular systole, just after the opening of a stenotic pulmonic valve, and is heard just after the QRS complex.

8. Unlike a PES, caused by pulmonic valve stenosis, an AES, caused by aortic valve stenosis, doesn't vary in intensity with respiration.

9. An AES is heard best near the heart's apex over the mitral area, near the heart's base over the aortic area, or over Erb's point. Its intensity is soft but may be equal to or greater than that of S_1. An AES has a short duration. It has a high pitch that's heard best using the diaphragm of the stethoscope, and it has a sharp, or clicklike, quality. Its timing is closely related to S_1. An AES occurs early in ventricular systole, just after the opening of a stenotic aortic valve, and is heard just after the QRS complex on the ECG waveform.

10. Certain characteristics help to distinguish an AES from other heart sounds. One characteristic is that an AES radiates more than a PES. Another characteristic is that a split S_1 heard over the mitral area is likely to be an M_1 and an AES.

   You can differentiate an AES from an S_4 by remembering that S_4 is heard best over the mitral area using the bell of the stethoscope and is commonly accompanied by a palpable, presystolic apical bulge. Also, an S_4 is intensified by maneuvers that increase left atrial pressures, such as brief exercise, squatting, or coughing. An AES is unaffected by these maneuvers.

11. A midsystolic click (MSC) occurs when the prolapsed mitral valve's leaflets and chordae tendineae tense. The anterior leaflet, the posterior leaflet, or both can prolapse.

12. An MSC is usually heard best over the tricuspid area and near the heart's apex over the mitral area. Its intensity is equal to or greater than that of S_1. An MSC has a short duration. It has a high pitch that's heard best using the diaphragm of the stethoscope, and it has a clicklike quality. The click's timing varies: It can occur in early systole, midsystole, or late systole. An MSC can be heard during the QT interval on the ECG waveform.

13. The timing of an MSC is affected by various maneuvers, such as having the patient stand or perform Valsalva's maneuver. Such maneuvers result in reduced left ventricular filling and cause the MSC to be heard closer to S_1. The MSC may even merge with S_1 and disappear completely. Increasing left ventricular volume by raising the legs from a recumbent position or squatting delays the click. This maneuver may also cause the prolapse not to occur and the MSC to be diminished or inaudible. Sometimes an MSC is accompanied by a late systolic crescendo murmur or the characteristic holosystolic murmur of mitral regurgitation. An MSC may also be caused by some extracardiac conditions, such as pleuropericardial adhesions.

7 MURMUR FUNDAMENTALS

1. Several clinically significant conditions — such as blood flowing at a high velocity through a partially obstructed opening, blood flowing from a higher pressure chamber to a lower pressure chamber, or a combination of these — can cause turbulent blood flow.

2. The terms used to describe a specific characteristic are determined primarily by the volume and speed of the jet of blood as it moves through the heart.

3. The seven characteristics used to describe murmurs are location, intensity, duration, pitch, quality, timing, and configuration.

4. A murmur's location is the anatomic area on the chest wall where the murmur is heard best and is usually also the murmur's point of maximum intensity. This area usually correlates with the underlying location of the valve that's responsible for producing the murmur.

5. The murmur's sounds may also be transmitted to the chamber or vessel where the turbulent blood flow occurs. This phenomenon, known as radiation, occurs because the direction of blood flow determines sound

transmission. Murmurs radiate in either forward or backward.

6. Intensity refers to the murmur's loudness.

7. To document a murmur's intensity, most health care professionals use a six-point graded scale, with 1 being the faintest intensity and 6 being the loudest. A grade 1 murmur is faint and is barely heard through the stethoscope. A grade 2 murmur is also faint but is usually heard as soon as the stethoscope is placed on the chest wall. A grade 3 murmur is easily heard and is described as moderately loud. A grade 4 murmur is loud and is usually associated with a palpable vibration known as a thrill. It also may radiate in the direction of blood flow. A grade 5 murmur is loud enough to be heard with only an edge of the stethoscope touching the chest wall; it's almost always accompanied by a thrill and radiation. A grade 6 murmur is so loud that it can be heard with the stethoscope close to, but not touching, the chest wall; it's always accompanied by a thrill and it radiates to distant structures.

8. Duration, the length of time the murmur is heard during systole or diastole, may be described as long or short.

9. A murmur's pitch, or frequency, varies from high to low.

10. A murmur's quality is determined by the combination of frequencies that produces the sound. It's typically described as harsh, rough, musical, scratchy, squeaky, rumbling, or blowing.

11. A murmur's timing refers to when the murmur occurs in the cardiac cycle. This means that the onset, duration, and end of the murmur are described in relation to systole and diastole. Murmurs are further classified according to their timing within the phases of the cardiac cycle. For example, a murmur can be described as holosystolic (present throughout systole); early systolic, midsystolic, or late systolic; or early diastolic, middiastolic, or late diastolic.

12. Configuration refers to the shape of a murmur's sound as recorded on a phonocardiogram. The configuration is usually defined by changes in the murmur's intensity during systole or diastole and is determined by blood flow pressure gradients.

13. A crescendo murmur is one that gradually increases in intensity as the pressure gradient increases. A de-

crescendo murmur is one that gradually decreases in intensity as the pressure gradient decreases. A crescendo-decrescendo murmur first increases in intensity as the pressure gradient increases, then decreases in intensity as the pressure gradient decreases; it's also known as a diamond-shaped murmur. Finally, a plateau-shaped murmur is one that's equal in intensity throughout the murmur.

8 SYSTOLIC MURMURS

1. Normally, as ventricular pressures rise at the beginning of systole, the mitral and tricuspid valves close. Then, for a brief time while the aortic and pulmonic valves are still closed during isovolumic contraction, ventricular pressures rise sharply. When the pressure in both ventricles is high enough, the aortic and pulmonic valves open, and blood is ejected from the ventricles into the aorta and the pulmonary artery. Normally functioning valves facilitate this unidirectional blood flow. However, aortic or pulmonic outlet abnormalities may generate forward systolic ejection murmurs. When the mitral or tricuspid valve is involved, backward, or regurgitant, murmurs may be heard during systole.

2. During ventricular systole, the rapid ejection of blood from the ventricles causes turbulent blood flow, which produces an innocent systolic ejection murmur (SEM).

3. A SEM is usually heard best along the left sternal border and sometimes over the aortic and mitral areas. Its intensity is usually soft (less than a grade 3/6), and its duration is short. It has a medium pitch that's heard best using the diaphragm of the stethoscope. The quality of a SEM varies. Its timing is early systolic; it ends well before a normal S_2 split. It's heard after the QRS complex on the ECG waveform. A SEM has a crescendo-decrescendo configuration.

4. A SEM's intensity can be increased by maneuvers that increase blood volume or ejection velocity, such as having the patient raise his legs from a recumbent position, exercise briefly, or cough a few times.

5. Right or left ventricular outflow obstructions may be supravalvular, valvular, or subvalvular. Regardless of location, the outflow obstruction causes turbulent blood flow, which produces a midsystolic ejection murmur. The murmur begins early in systole, after S_1

and the opening of the diseased pulmonic or aortic valve. It ends before the S_2 closure component of the diseased pulmonic or aortic valve. This murmur typically has a crescendo-decrescendo configuration that peaks in intensity in early systole, midsystole, or late systole, depending on the severity of the obstruction.

6. The supravalvular pulmonic stenosis murmur is usually heard over much of the thorax. Its intensity and duration are variable. It has a medium pitch that's heard best using the diaphragm of the stethoscope, and it has a harsh quality. Its timing is systolic: It begins after S_1 and ends before a normal S_2 split. On the ECG waveform, the murmur begins just after the QRS complex begins and ends just before the T wave ends. It has a crescendo-decrescendo configuration that's occasionally continuous.

7. In mild pulmonic valvular stenosis, S_1 is normal. The murmur begins after S_1 with a right-sided pulmonic ejection sound (PES) as the pulmonic valve abruptly stops opening. The murmur intensifies after the PES, peaks in midsystole, and then begins to fade. It ends before S_2. The intensity of P_2 is normal.

   In severe pulmonic valvular stenosis, the pressure gradient across the pulmonic valve increases. An increased pressure gradient causes the PES to be heard earlier; it may even fuse with S_1. Right ventricular ejection time is also prolonged. Consequently, the murmur has a longer crescendo and the intensity peaks later in systole. It continues throughout A_2 but ends before P_2. Prolonged right ventricular ejection time also causes a delayed P_2, creating a wide S_2 split. Usually, as the stenosis becomes more severe, the murmur's duration lengthens, and its configuration becomes more asymmetrical. Consequently, P_2 is delayed even more and its intensity is decreased.

8. The pulmonic valvular stenosis murmur is heard best near the heart's base over the pulmonic area. It commonly radiates toward the left neck or the left shoulder. The murmur's intensity is usually soft, but it will become louder as the stenosis becomes more severe. Its duration is short, but this too will increase as the stenosis worsens. It has a medium pitch that's heard best using the diaphragm of the stethoscope, and it has a harsh quality. The murmur's timing is midsystolic: It ends before a normal S_2 split. In the ECG waveform, it

begins after the QRS complex and ends before the T wave ends. It has a crescendo-decrescendo configuration. A mild pulmonic valvular stenosis murmur is shaped like a diamond; a severe pulmonic valvular stenosis murmur is shaped like a kite.

9. The subvalvular pulmonic stenosis murmur is usually heard best over the pulmonic area and over Erb's point; it commonly radiates toward the left side of the neck, the left shoulder, or both. Its intensity is usually soft but becomes louder as the stenosis worsens. Its duration, which is short, also increases as the stenosis worsens. The murmur has a medium pitch that's heard best using the diaphragm of the stethoscope, and it has a harsh quality. Its timing is midsystolic: It starts after S_1 and ends before a normal S_2 split. In the ECG waveform, the murmur begins after the QRS complex and ends before the end of the T wave. This murmur isn't initiated by a pulmonic ejection sound. It has a crescendo-decrescendo configuration.

10. The supravalvular aortic stenosis murmur is usually heard best near the heart's base over the right first intercostal space, over the aortic area, and over the suprasternal notch. It may radiate toward the right side of the neck, the right shoulder, or both. Its intensity, typically a 3/6 to 4/6, decreases in patients with left-sided heart failure. Its duration increases as the stenosis worsens. It has a medium pitch that can be heard equally well using the diaphragm or bell of the stethoscope. The murmur has a rough quality. Its timing is midsystolic: It ends before a normal S_2 split. It isn't associated with an aortic ejection sound. On the ECG waveform, it begins after the QRS complex and ends before the end of the T wave. The murmur has a crescendo-decrescendo configuration.

11. The aortic valvular stenosis murmur is usually heard best near the heart's base over the aortic area, over Erb's point, near the heart's apex over the mitral area, or over the suprasternal notch. The murmur may radiate toward the right side of the neck, the right shoulder, or both. A thrill may be palpable over the aortic area and neck. The intensity, typically a grade 3/6 to 4/6, decreases in patients with left-sided heart failure. The murmur's duration increases as stenosis worsens.

     The aortic valvular stenosis murmur has a medium pitch that's heard equally well using the diaphragm or bell of the stethoscope. The murmur's quality is rough and may become harsher and louder as stenosis worsens. Its timing is midsystolic. An aortic ejection sound (AES), when present, is heard shortly after S_1; the AES is followed by the murmur, which ends before a normal S_2 split. In the ECG waveform, the murmur begins after the QRS complex and ends before the end of the T wave. It has a crescendo-decrescendo configuration.

12. A subvalvular aortic outflow obstruction may be caused by a congenital fibrous ring, or it may be acquired from hypertrophic obstructive cardiomyopathy.

13. The subvalvular aortic stenosis murmur is usually heard best near the heart's apex over the mitral and tricuspid areas. It doesn't usually radiate toward the base, right side of the neck, or the right shoulder. Its intensity, typically a grade 3/6 to 4/6, will increase as stenosis worsens; its duration varies. The murmur has a medium pitch heard equally well with the bell or diaphragm of the stethoscope. Its quality can be harsh or rough. This murmur's timing is midsystolic: It peaks in midsystole and ends before a normal S_2 split, a delayed A_2, or a paradoxical S_2 split. On the ECG waveform, it begins after the QRS complex and ends before the end of the T wave. The murmur has a crescendo-decrescendo configuration.

14. An abnormality of either the mitral or tricuspid valve may result in backward turbulent blood flow during systole. This means that blood moves in a direction opposite that of the normal unidirectional flow pattern. Blood regurgitates through a defective, incompetent mitral or tricuspid valve into the left or right atrium.

15. The tricuspid regurgitation murmur is usually heard best over the tricuspid area; in some patients, it's heard only during inspiration. The murmur may radiate to the right of the sternum. Its usually soft intensity may increase during deep inspiration. The murmur's duration is long. It has a medium pitch that's heard best using the diaphragm of the stethoscope, and it has a scratchy or blowing quality. The murmur's timing is systolic: It lasts from S_1 to P_2. On the ECG waveform, the murmur begins just after the QRS complex and ends after the T wave. It's holosystolic and plateau shaped.

16. To enhance the murmur, have the patient breathe through his mouth slowly, quietly, and more deeply while sitting or standing. Caution the patient not to hold his breath because this negates the maneuver's effect. The murmur is louder during inspiration. In some patients, this is the only time the murmur is audible.

17. An incompetent mitral valve causes backward blood flow through the incompetent mitral valve during systole. The increased pressure at the aortic valve facilitates regurgitation of blood through the incompetent mitral valve into the left atrium, producing the mitral regurgitation murmur.

18. The holosystolic mitral regurgitation murmur is usually heard best near the heart's apex over the mitral area. The sound may radiate to the axillae or posteriorly over the lung bases. Its intensity varies and usually is unaffected by respiration; however, it may be somewhat diminished during inspiration. The murmur has a long duration. It has a medium to high pitch that's heard best using the diaphragm of the stethoscope. The murmur may be accompanied by a systolic apical thrill. Its quality is blowing, and its timing is systolic — from S_1 to S_2. On the ECG waveform, it's heard from just after the QRS complex to the end of the T wave. The murmur is holosystolic and plateau shaped.

19. The acute mitral regurgitation murmur is usually heard best near the heart's apex over the mitral area. Its intensity is usually loud (a grade 4/6 to 6/6 if the murmur results from rupture of the chordae tendineae). It's accompanied by a systolic thrill. The murmur's duration is medium long. It has a high pitch that's heard best using the diaphragm of the stethoscope, and it has a musical quality. Its timing is systolic: It begins with M_1 and ends with or before A_2. On the ECG waveform, the murmur begins just after the QRS complex and ends just after the T wave. It's commonly holosystolic and wedge shaped (the wedge shape has a steeper decrescendo configuration).

20. The mitral valve prolapse murmur is usually heard best near the heart's apex over the mitral area. Its intensity is usually soft (a grade 2/6 to 3/6), and its duration is short. It has a high pitch heard best with the diaphragm of the stethoscope. The murmur has a musical quality; when loud, it's sometimes described as a

whoop or honk. Its timing is usually late systolic but can sometimes be holosystolic. In the ECG waveform, the murmur coincides with the T wave and ends just after the T wave. It has a crescendo or crescendo–decrescendo configuration.

21. Having the patient stand decreases left ventricular volume, which causes the mitral valve prolapse murmur to be heard earlier in systole, to be louder, and to last longer. Having the patient squat increases left ventricular volume, which causes the murmur to be heard later in systole.

9 DIASTOLIC MURMURS

1. At the end of ventricular systole and the beginning of diastole, the aortic and pulmonic valves close. During this time, the A_2-P_2 interval can usually be auscultated. After a brief period of isovolumic relaxation, the mitral and tricuspid valves open, and blood flows from the atria into the ventricles. During this early filling period, an S_3 can sometimes be heard in healthy individuals younger than age 20. Late in diastole, the atria contract, increasing blood flow into the ventricles. Occasionally, an S_4 can be heard during this late filling period. Except for these brief heart sounds, diastole is normally silent. When diastolic murmurs occur, they're heard between S_2 and S_1 or between the end of the T wave and the beginning of the QRS complex on the ECG waveform.

   The regurgitation of blood through the aortic and pulmonic valves may cause diastolic murmurs. Because both valves close at the beginning of diastole, murmurs produced by dysfunctional aortic and pulmonic valves begin early in diastole, immediately after the affected valve closes.

   Mitral and tricuspid valve stenosis and conditions that produce turbulent blood flow across normal mitral or tricuspid valves also cause diastolic murmurs. Because these valves open after a period of isovolumic relaxation, these murmurs are heard during middiastole.

2. The early diastolic aortic regurgitation murmur is usually heard best near the heart's base over the aortic and pulmonic areas, over Erb's point, and near the heart's apex over the mitral area. Because of its usually soft intensity, it's heard best in a quiet environment. The

murmur can last throughout most of diastole. It has a high pitch that's heard best using the diaphragm of the stethoscope, and it has a blowing or musical quality. Its timing is diastolic, beginning with A_2. On the ECG waveform, the murmur begins after the T wave ends and ends just before the QRS complex. It has a decrescendo configuration.

3. This murmur can be enhanced by having the patient sit down, lean forward, and hold his breath after expiration or perform maneuvers that increase aortic diastolic pressure, such as squatting or performing handgrip exercises.

4. The Austin Flint murmur is usually heard best near the heart's apex over the mitral area. Its intensity is usually soft. It has a low pitch that's heard best using the bell of the stethoscope. The murmur has a rumbling quality. Its timing is confined to middiastole and presystole. It's heard just before the QRS complex on the ECG waveform. The murmur has a presystolic crescendo configuration; its middiastolic component has a crescendo-decrescendo configuration.

5. The Graham Steell murmur secondary to pulmonary hypertension is usually heard best along the left sternal border over the third and fourth intercostal spaces. It isn't transmitted to the right sternum. Its intensity is usually loud, and its duration varies. The murmur has a high pitch that's heard best using the diaphragm of the stethoscope, and it has a blowing quality. Its timing is early diastolic, beginning with a loud P_2. Sometimes an ejection sound can be heard. The murmur is heard after the end of the T wave in the ECG waveform. It has a decrescendo configuration.

6. The Graham Steell murmur is intensified during inspiration.

7. The normal pressure pulmonic valve murmur is usually heard best along the left sternal border over the third and fourth intercostal spaces. It isn't transmitted to the right sternum. Its intensity is soft and its duration brief. The murmur has a low pitch that's heard best using the bell of the stethoscope. It has a rumbling quality and its timing is early diastolic to middiastolic. It begins shortly after P_2 is heard. On the ECG waveform, it begins after the T wave and ends before the P wave. The murmur has a crescendo-decrescendo configuration.

8. The normal pressure pulmonic valve murmur is intensified during inspiration.

9. Scarring and calcification of the valve create a funnel-shaped opening between the left atrium and left ventricle. The mitral stenosis murmur starts with an opening snap (OS) after A_2 and isovolumetric relaxation; this OS is an important diagnostic feature of mitral stenosis. The murmur is produced by rapid, turbulent blood flow through a rigid, narrowed mitral valve opening. Turbulence increases just before systole; this gives the murmur its characteristic presystolic crescendo.

10. The mitral stenosis murmur is usually heard best near the heart's apex over the mitral area. Its intensity and duration vary. It has a low pitch that's heard best using the bell of the stethoscope and a rumbling quality that has been compared to thunder. The timing of a mitral stenosis murmur is diastolic: It begins after A_2 with an OS and ends with a loud M_1. However, stenosis severity affects the murmur's duration. On the ECG waveform, the murmur begins just after the T wave and ends during the QRS complex. It has a crescendo-decrescendo configuration. In patients with normal sinus rhythm, the presystolic component has a crescendo configuration.

11. The mitral stenosis murmur is heard best with the patient in a partial left lateral recumbent position. It can be intensified by maneuvers that increase cardiac output, such as having the patient exercise for a few minutes, raise his legs from a recumbent position, or cough several times.

12. The tricuspid stenosis murmur is usually heard best over the tricuspid area. Its usually soft intensity increases during inspiration and fades or disappears during expiration. The murmur's duration varies. It has a low pitch that's heard best using the bell of the stethoscope, while the patient is in the partial left lateral recumbent position. The murmur has a rumbling quality. Its timing is middiastolic to late diastolic: It ends just before S_1. An opening snap may be heard. In the ECG waveform, the murmur begins just after the T wave and ends just before the QRS complex. In patients with normal sinus rhythm, it has a late diastolic crescendo or crescendo-decrescendo configuration.

10 CONTINUOUS MURMURS

1. Continuous murmurs are generated by rapid blood flow through arteries or veins or by shunting. Shunting occurs when an abnormal communication is created between the high-pressure arterial system and the low-pressure venous system. The murmurs begin in systole and persist, without interruption, through S_2 into diastole.

2. The normal cervical venous hum is caused by rapid downward blood flow through the jugular veins in the lower part of the neck.

3. The cervical venous hum murmur is heard best over the right supraclavicular fossa when the patient is sitting and his head is turned to the left. It has a faint intensity that increases during diastole. The murmur has a long duration: It lasts throughout the cardiac cycle. Its high pitch is heard best using the diaphragm of the stethoscope. Its quality is soft and its timing continuous. The hum is louder between S_2 and S_1. The murmur has a plateau shape in systole and a crescendo-decrescendo configuration in diastole.

4. The patent ductus arteriosus (PDA) murmur is heard best over the left first intercostal space, over the pulmonic area, and below the left clavicle. Its intensity is faint when the murmur is limited to this area. If the murmur is loud, the systolic component may be heard along the left sternal border and sometimes over the mitral area. A loud PDA murmur may radiate to the back between the scapulae. The PDA murmur has a long duration. Its intensity varies, roughly correlating with the ductus size; typically, the murmur reaches maximum intensity late in systole and then fades during diastole. The PDA murmur has a high pitch that's heard equally well using the bell or diaphragm of the stethoscope. It has a rough, machinery-like quality, and its timing is continuous. It begins with — or shortly after — a normal S_1 and disappears just before the next S_1. Because the murmur peaks in late systole, S_2 may be difficult to hear. S_2 may be paradoxically split if left ventricular ejection time is prolonged. An S_3 may be heard over the mitral area. The murmur has a crescendo-decrescendo configuration.

5. Exercise can increase the intensity and duration of a PDA murmur.

6. To differentiate between a PDA murmur and a cervical venous hum murmur, remember that the hum is loudest above the clavicle, is usually heard better on the right, and can be obliterated by pressing on the jugular vein or placing the patient in a supine position. The venous hum is truly continuous and is usually louder during diastole.

11 OTHER AUSCULTATORY SOUNDS

1. The aortic and mitral valves are most commonly replaced.
2. The four types of prosthetic valves used to replace human valves are the ball-in-cage, the tilting-disk, the bileaflet, and the porcine valves.
3. The murmur generated by an aortic ball-in-cage valve prosthesis is usually heard best near the apex over the mitral and aortic areas and along the left sternal border. Its intensity is usually loud and easy to hear, and its duration varies. It has a medium pitch that's heard best using the diaphragm of the stethoscope. The murmur has a crunchy, harsh quality. Its timing is midsystolic. The interval between the aortic closing click and P_2 is similar to the normal A_2-P_2 interval, and it normally widens during inspiration. The murmur has a crescendo-decrescendo configuration.
4. Dysfunction of an aortic ball-in-cage valve prosthesis commonly causes the aortic opening click to become soft or absent; the aortic closing click may be absent, too. A diastolic murmur and a long systolic ejection murmur may appear.
5. The murmur generated by an aortic tilting-disk valve prosthesis is usually heard best near the apex over the mitral and aortic areas and along the left sternal border. Its intensity is usually soft (grade 2/6) and its duration short. It has a medium pitch that's heard best using the diaphragm of the stethoscope. The murmur has a rough or harsh quality. Its timing is systolic. The interval between the aortic closing click and P_2 is similar to the normal A_2-P_2 interval, and it widens during inspiration. The murmur has a crescendo-decrescendo configuration.
6. Dysfunction of an aortic tilting-disk valve prosthesis commonly causes some or all of the following changes: The aortic closing click may be absent, a di-

astolic murmur may be heard, or a longer systolic ejection murmur may appear.

7. The murmur generated by an aortic bileaflet valve prosthesis is usually heard best near the apex over the mitral and aortic areas and along the left sternal border. Its intensity is usually soft (grade 2/6) and its duration short. It has a medium pitch that's heard best using the diaphragm of the stethoscope. The murmur has a rough or harsh quality, and its timing is systolic. The interval between the aortic closing click and P_2 is similar to the normal A_2-P_2 interval, and it widens during inspiration. The murmur has a crescendo-decrescendo configuration.

8. Dysfunction of the aortic bileaflet valve prosthesis may cause the aortic closing click to disappear, a diastolic murmur to appear, or both. A longer systolic ejection murmur also may appear.

9. The murmur generated by an aortic porcine valve prosthesis is usually heard best near the apex over the mitral and aortic areas and along the left sternal border. Its intensity is usually soft (grade 2/6) and its duration short. It has a medium pitch that's heard best using the diaphragm of the stethoscope. The murmur has a rough or harsh quality, and its timing is systolic. The interval between the aortic closing sound and P_2 is similar to the normal A_2-P_2 interval, and it widens during inspiration. The murmur has a crescendo-decrescendo configuration.

10. Dysfunction of an aortic porcine valve prosthesis may cause the aortic closing click to be diminished, a diastolic murmur to appear, or both. A longer systolic ejection murmur also may appear.

11. With a mitral ball-in-cage valve, M_1 is loud and higher in frequency. A mitral opening click follows a normal S_2.

12. Dysfunction of a mitral ball-in-cage valve prosthesis may cause some or all of the following changes: The sounds normally associated with the valve may vary in intensity in a patient with normal sinus rhythm; a short interval between A_2 and the mitral opening click may occur, indicating high left atrial pressure; a holosystolic murmur may occur, indicating mitral regurgitation; or the intensity and duration of the diastolic murmur may change, indicating mitral obstruction.

13. A normally functioning mitral tilting-disk valve prosthesis replaces M_1 with a mitral closing click. This closing click is always distinctly audible and high pitched; its maximum intensity is located near the apex over the mitral area.

14. Dysfunction of a mitral tilting-disk valve prosthesis may cause some or all of the following changes: The mitral closing click may be absent, a new diastolic murmur may develop, a previously auscultated diastolic murmur may intensify, or a holosystolic mitral regurgitation murmur may appear.

15. In the systolic murmur generated by the mitral bileaflet valve, the first sound is replaced by a higher-pitched sound. The mitral opening click, though rarely heard, follows a normal S_2.

16. Dysfunction of a mitral bileaflet valve prosthesis may cause a holosystolic murmur, a new diastolic murmur, or both to appear.

17. The systolic murmur generated by a mitral porcine valve prosthesis is usually heard best near the apex over the mitral area and sounds like a normal M_1.

18. Dysfunction of a mitral porcine valve prosthesis may cause a holosystolic mitral regurgitation murmur, a diastolic rumble associated with mitral stenosis, or both to appear.

19. When inflamed pericardial surfaces rub together, they produce characteristic high-pitched friction noises known as pericardial friction rubs.

20. A pericardial friction rub is usually heard best—and is sometimes palpable—over the tricuspid and xyphoid areas. It's usually loud and may get louder during inspiration. The rub has a high pitch that's heard best using the diaphragm of the stethoscope, and it has a grating or scratchy quality. It's usually heard during each cardiac cycle. It has a systolic component and an early and late diastolic component. The diastolic component may last for only a few hours.

21. You can easily differentiate a pericardial friction rub from a pleural friction rub by having the patient hold his breath. When he does so, a pericardial friction rub will persist, but a pleural friction rub will become inaudible.

22. Heart movements can displace air that is present in the mediastinum; this displacement produces crunchy

noises—known as mediastinal crunch—that may occur randomly or in a consistent pattern.

23. The noises produced by mediastinal crunch are usually heard best along the left sternal border with the patient in the sitting position. The noises have a crunching quality, which may become louder during inspiration.

Breath sounds

12 THE RESPIRATORY SYSTEM AND AUSCULTATION

1. The respiratory system's primary function is the exchange of oxygen (O_2) and carbon dioxide (CO_2) between the alveoli and the pulmonary circulation.
2. The upper airway consists of the nasal cavities and the pharynx.
3. The pharynx is divided into three sections: the nasopharynx, the oropharynx, and the hypopharynx.
4. The lower airway is called the tracheobronchial tree. It begins at the cricoid cartilage of the larynx and ends at the distal bronchioles, which open into the alveoli.
5. Anteriorly, the trachea begins just below the larynx and is located within the thoracic cavity, beneath the upper two-thirds of the sternum. Posteriorly, it begins at the level of the sixth cervical or first thoracic vertebra and extends to about the fifth thoracic vertebra.
6. The mainstem bronchi enter the lungs at the hila, where the lung tissue attaches to the mediastinum. The bronchi course downward and immediately divide into lobar bronchi, which subdivide into segmental bronchi. The airways continue to systematically branch, narrow, shorten, and increase in number toward the lung periphery. They divide about 25 times before ultimately opening into the terminal, or respiratory, bronchioles. The terminal bronchioles, which have a few alveoli budding from their walls, branch into the alveolar ducts, which are completely lined with alveoli.
7. Anatomic dead space is the about 150 cc of air that remains in the conducting airways during each breath. This air isn't involved in O_2-CO_2 exchange.
8. The mucous and bronchial glands located in the bronchial walls secrete a serous mucus blanket onto the inside airway walls. This secreted mucus forms a

thin, sticky layer that lies on top of ciliated columnar epithelial cells, which line nearly all of the respiratory tract between the larynx and the terminal bronchioles. The mucus layer traps foreign particles in the inspired air before they can enter the lungs. The cilia move in a wavelike motion, propelling the sticky mucus layer toward the pharynx. Together, the mucus layer and cilia protect the lungs, acting as a major defense mechanism against inhaled particles.

9. Each lung is divided into lobes. The right lung has three lobes: the upper, middle, and lower. The left lung has only an upper and a lower lobe.

10. The lung's apex (superior aspect) is located near the clavicle, and its base (inferior aspect) is located near the diaphragm.

11. Alveoli contain three cell types. Type 1, the largest, is the lining cell and accounts for 95% of the alveolar surface area. The type 2 cell is believed to produce surfactant, a mixture of phospholipids. The macrophage, the third cell type, acts as a phagocytic defense mechanism against infection.

12. Surfactant is a mixture of phospholipids found in the fluid layer that covers the alveolar surface. The surfactant mixes with pulmonary fluids and maintains alveolar stability by lowering the alveolar surface tension, thus preventing the air-filled alveoli from collapsing at low volumes from the weight of the capillary blood.

13. The parasympathetic nervous system in the larynx, trachea, and bronchi includes vagal nerve fibers and irritant receptors mediated by the neurotransmitter acetylcholine. Parasympathetic stimulation precipitates bronchoconstriction, coughing, and mucus discharge from the bronchial glands. Irritant receptors within the airways are believed to be responsible for increasing ventilation and precipitating coughing.

    Sympathetic innervation of the lower airways usually arises from the thoracic nerves T_1 through T_5 and is mediated by the postganglionic hormone transmitter norepinephrine. Sympathetic stimulation of $beta_2$-adrenergic receptors throughout the lower airways facilitates smooth-muscle relaxation, resulting in bronchodilation.

14. The pulmonary circulation begins as the pulmonary artery leaves the right ventricle. The pulmonary artery divides into segmental arteries, which subdivide, end-

ing in the pulmonary capillary bed surrounding the alveoli. The capillaries then join to form venules, which converge to form veins, then pulmonary veins. The pulmonary veins end in the left atrium.

15. The bronchial circulation usually originates in the aorta and branches along the bronchi to provide oxygen and nutrients to the lung tissue and to clear metabolic wastes from the airways and other lung tissues. Oxygenated blood is pumped from the aorta to the tracheobronchial tree and travels through arterioles along the airways. Deoxygenated blood from the bronchial circulation is returned to the left atrium, accounting for the normal 2% to 3% right-to-left shunt.

16. The chest wall's inner surface and the lungs' outer surfaces are joined by thin, membranous tissues called the pleurae. The parietal pleura covers the chest wall's inner surface, and the visceral pleura covers the lungs' outer surfaces.

17. The thorax is the bony structure that protects the vital organs (the heart and lungs) and permits chest expansion during inspiration.

18. The diaphragm, the primary muscle of respiration, consists of two dome-shaped hemidiaphragms that are attached to the lower edge of the rib cage. The diaphragm forms the inferior "floor" of the thorax. During inspiration, the diaphragm flattens and descends toward the abdomen. During expiration, it relaxes and ascends to its resting, dome-shaped configuration.

19. The external intercostal muscles, located between the ribs and innervated by the intercostal nerves, contract during inspiration to stabilize, elevate, and expand the rib cage.

20. The accessory muscles of respiration — the sternocleidomastoid, scalene, trapezius, and rhomboid muscles — are involved in labored or forceful breathing. The sternocleidomastoid muscle elevates the sternum, increasing the anterior-to-posterior chest diameter. The scalene muscles elevate and fix the first two ribs. The trapezius and rhomboid muscles are activated during respiration but probably have no effect on forceful inspiration. The contraction of external intercostal muscles may also be obvious in extremely labored breathing.

21. During inspiration, the diaphragm and external intercostal muscles are activated. Diaphragmatic contrac-

tion flattens the domed diaphragm and expands the lower rib cage, forcing the abdominal contents downward and out, thus increasing the longitudinal lung size. External intercostal muscle contraction stabilizes the rib cage and moves it outward and upward. The posterior cricoarytenoid muscle contracts to open the glottis. As the thorax expands, intrapleural pressures become subatmospheric, resulting in lung expansion and decreased intrapulmonary pressures. Inspired air flows from higher atmospheric pressure into the airways. At the end of inspiration, diaphragmatic movement declines and the air inflow gradually slows.

During expiration, the thorax and the elastic recoil force of the lungs return to their resting positions, increasing intrapleural pressures. This increased pressure forces air to flow out of the lower and upper airways. At the end of expiration, during quiet breathing, all muscles are relaxed and the diaphragm has returned to its resting position.

22. During normal breathing in the upright position, air movement (ventilation) is greatest in the lung bases or dependent lung regions. Because of the pull of gravity, the alveoli in the lung bases collapse to a smaller size during expiration than do the alveoli in the lung apices. During inspiration, the alveoli in the lung bases open more easily, causing ventilation in the bases to be greater than in the apices.

Blood perfusion in the lungs is also gravity dependent; the dependent lung regions receive a larger proportion of the cardiac output than the nondependent regions.

23. Postural changes affect the distribution of ventilation and blood flow throughout the lungs. In the supine position, the apices and bases receive equal amounts of air and blood flow. In the lateral position, dependent lung regions receive more air and blood flow than nondependent regions.

24. The tracheobronchial tree's dimensions and shape affect the airflow pattern as follows: During inspiration, air velocity and the airflow rate decrease as the cross-sectional airway area increases. Consequently, the larger airways conduct the same volume of air as do the more numerous smaller airways. This means that inspired air travels 100 times faster in the trachea than it does in the terminal bronchi. Beyond the terminal

bronchi, the forward movement of air is minimal. In the terminal bronchioles, diffusion is the primary method of air movement.

25. The lung apices extend 1¼" to 1½" (2 to 4 cm) above the clavicles. The trachea bifurcates at the level of the sternal angle, the junction between the manubrium and the body of the sternum. The ribs and intercostal spaces provide precise horizontal landmarks to describe the location of breath sounds; the second rib and second intercostal space serve as reference points.

    Vertical landmark lines on the anterior chest wall surface include the midclavicular lines and the midsternal line. The right and left midclavicular lines extend downward from the center of each clavicle. The midsternal line bisects the sternum. The right lung base crosses the sixth rib at the right midclavicular line, and the left lung base crosses the seventh rib at the left midclavicular line.

26. The anterior axillary line extends downward from the anterior axillary fold, the midaxillary line extends downward from the apex of the axilla, and the posterior axillary line extends downward from the posterior axillary fold.

27. Bony structures underlying the posterior chest wall surface also provide landmarks to locate breath sounds. The scapulae's inferior borders, located at about the same level as the seventh rib, serve as reference points. The numbered thoracic vertebrae provide horizontal landmarks. Landmark lines on the posterior chest wall surface provide vertical reference points. The midscapular lines extend downward from the inferior angle of each scapula, and the vertebral line extends downward over the vertebrae.

28. The diaphragm attenuates low-pitched sound frequencies, so it's used to listen for high-pitched breath sounds; the bell attenuates high-pitched sound frequencies, so it's used to listen for low-pitched breath sounds.

29. During auscultation, press the diaphragm of the stethoscope firmly against the patient's chest wall over the intercostal spaces. Try not to listen directly over bone; never listen through clothing, which impedes or alters sound transmission.

30. If the patient is alert and healthy, begin auscultation with the patient sitting upright and leaning slightly for-

ward with his hands grasping his elbows to separate
the scapulae. Position yourself behind the patient. Ask
the patient to breathe through an open mouth, slightly
deeper and faster than usual, through several respiratory cycles. Place the diaphragm of the stethoscope over
the left lung apex, and listen for at least one complete
respiratory cycle. Then move the diaphragm to the
same site over the right lung. Compare the breath
sounds heard over these same locations. Continue in
this manner, making contralateral comparisons at each
auscultatory site.

13 INTRODUCTION TO BREATH SOUNDS

1. Breath sounds are produced by airflow patterns, by associated pressure changes within the airways, and by
   solid-tissue vibrations.

2. Sounds heard over the trachea and large airways are
   characterized as loud and tubular (hollow) with a long
   expiratory phase. These sounds are thought to reflect
   the turbulent airflow patterns in the first divisions of
   the large airways. Sounds heard over other chest wall
   areas are softer and have a shorter expiratory phase.

3. An airway's stability depends on the interaction among
   the elastic properties of the lung tissue, the intrapulmonary airways, and the pressures—both internal and
   external—exerted on the airways. During the respiratory cycle, a pressure gradient exists across the airway
   wall and between the trachea and alveoli. As the chest
   expands during inspiration, intrapleural pressure falls.
   At the same time, the lung's elastic recoil increases,
   exerting traction on airway walls. This traction increases the airways' diameter and decreases the resistance
   to airflow. The pressure gradient between the alveolar
   and atmospheric pressures increases; this drives air
   through the airways toward the alveoli.

4. Intrapleural pressure and alveolar elastic recoil pressures drive expiratory airflow. At the beginning of expiration, intrapleural pressure and the alveolar recoil
   force cause the alveolar pressures to rise and push the
   air through the airways and out through the mouth. Air
   pressures fall as the air flows through the airways from
   the alveoli to the mouth. A point is reached, particularly during forceful expiration, when intrapleural and intrapulmonary pressures are equal; this normally occurs

in the segmental bronchi and is called the equal pressure point. Downstream from this point, pressures outside of the airways can compress and further narrow the airways, causing increased resistance to airflow.

5. Variations in airflow patterns are affected by the intricate network of branching airways, the various airway diameters, and the potential for irregular airway wall surfaces within the tracheobronchial tree.

6. During rapid or turbulent airflow, air molecules move randomly, colliding against airway walls and each other. They may move across or against the general direction of airflow. The colliding air molecules produce rapid pressure changes within the airway, and these rapid pressure changes produce sounds. Turbulent airflow occurs in the trachea, mainstem bronchi, and other larger airways.

7. As airflow is forced to change directions abruptly in the branching airways, the airstream separates into layers and moves at different velocities. The shearing force of high-velocity airstreams, along with slower airstreams, precipitates opposing circular airflows, or vortices. This airflow pattern generates sounds as the flow of air carries the vortices downstream.

8. In the terminal or small airways and respiratory bronchioles, airflow is slow and laminar. No abrupt changes in pressure or airway wall movements occur to generate sound. Consequently, air movement in these areas produces no sound.

9. Frequency, measured in hertz, refers to the number of vibrations occurring per unit of time. The term *pitch* is used to describe sound frequency. High-pitched sounds have higher frequencies; low-pitched sounds, lower frequencies. Intensity is the loudness or softness of the vibrations producing breath sounds. It can be measured electronically by recording amplitude. Intensity is affected by the type of vibrating structure producing the sound, the distance the sound travels, and the transmission pathway. The duration of the vibrations that produce breath sounds can be measured in milliseconds. However, the human ear perceives sounds during auscultation as long or short and as continuous or discontinuous.

10. Body structures or cavities that produce resonance will selectively transmit or amplify breath sounds that are similar in frequency and will absorb or dampen those

with other frequencies. Breath sounds arising from the same location in the lung have a higher pitch when heard at the mouth or over the trachea than when heard over the chest wall. This change in pitch occurs because high-pitched breath sounds are absorbed as they're transmitted through the lungs and thorax.

11. Breath and voice sounds are normally dampened when they're transmitted through air, fluid, and tissue. However, when these sounds are transmitted sequentially through media with different acoustical properties, the sounds' vibrations are filtered and reflected at the interface of the different media. The amount of filtering that takes place depends on the media's impedance (resistance to sound transmission).

    When two media are matched according to acoustical properties, sound is transmitted effectively. For example, consolidated areas enhance the transmission of breath sounds to the chest wall. This happens because the consolidating substance, an inflammatory exudate, collects in alveoli, replacing air with dense lung tissue. Consequently, the fluid-filled, airless lung tissue and the chest wall are acoustically well matched, and breath sounds are transmitted more easily.

12. The transmission of breath and voice sounds through either air or fluid between inflated lung segments and the chest wall results in an impedance mismatch. The vibrations of the sounds are significantly filtered and reflected back to their source by the pleurae.

13. The tracheal and mainstem bronchi sounds, produced by turbulent airflow, are loud and can be heard throughout inspiration and expiration over the trachea and mainstem bronchi. Normal breath sounds heard over other chest wall areas are faint and can be heard throughout inspiration and at the beginning of expiration.

14. Bronchovesicular breath sounds represent a third classification of normal sounds. These sounds are heard over areas between the mainstem bronchi and the smaller airways; their pitch and duration are midway between those of tracheal and mainstem bronchi breath sounds and normal breath sounds heard over other chest wall areas. Bronchovesicular sounds are heard during inspiration and expiration for equal amounts of time.

15. Voice sounds auscultated over the chest wall are called bronchophony. In a healthy individual, bronchophony is similar to voice sounds heard through the neck. Voice sounds transmitted through the chest wall that have selectively amplified higher frequencies are called egophony. High-pitched whispered sounds transmitted through airless, consolidated lung tissue are called whispered pectoriloquy.

16. Adventitious sounds are added sounds that are heard with abnormal breath sounds. They're classified into two categories: crackles and wheezes.

17. Location, intensity, duration, and pitch are the four characteristics used to describe breath sounds.

18. Sound intensity is described subjectively, using such terms as *loud, soft, absent, diminished,* or *distant.*

19. Sound duration refers to the sound's timing within the respiratory cycle — that is, whether it's heard during inspiration, expiration, or both. Timing can be described as early, late, or throughout.

20. Sound pitch is usually described as high or low.

14 BREATH SOUNDS HEARD IN HEALTHY INDIVIDUALS

1. Normal breath sounds are broadly defined as breathing heard through the chest wall of a healthy individual.

2. Three variables affect the normal sounds heard during auscultation: the distance between the source of the sound and the chest wall, the path of sound transmission, and the sound's location.

3. Normal breath sounds are produced by air flow patterns, by associated pressure changes within the airways, and by solid-tissue vibrations.

4. Tracheal and mainstem bronchi breath sounds are heard over the chest wall on either side of the sternum from the second intercostal space to the fourth intercostal space anteriorly and along the vertebral column from the third intercostal space to the sixth intercostal space posteriorly. These sounds have a variable pitch and a loud intensity that's heard best using the diaphragm of the stethoscope. They're heard throughout inspiration and expiration.

5. These normal breath sounds are heard throughout the chest anteriorly, posteriorly, and laterally. Their duration varies, depending on their location, but inspiration is usually followed immediately by a shorter expira-

tion. The sounds have a low pitch that can be heard using either the diaphragm or bell of the stethoscope.

6. Variations in intensity are usually related to airflow patterns and to the distribution of ventilation throughout the lungs. During inspiration, with the patient in the upright position, air flows initially into the lung apices, then into the lung bases (the dependent areas) after the small airways reopen; therefore, intensity will probably be greater in the apices than in the bases. During normal breathing, this change in intensity may not always be audible.

7. Intensity may increase during systole because ventricular contraction allows the surrounding lung tissue to expand more fully, increasing turbulent airflow to that region and intensifying the inspiratory sounds. Conversely, intensity may decrease during diastole because ventricular expansion of the heart compresses adjacent lung tissue, reducing airflow to that region.

8. Normal breath sounds heard at the mouth are thought to be produced by turbulent airflow occurring below the glottis and before the terminal airways. In healthy individuals, breath sounds at the mouth provide baseline data that can be useful later when evaluating noisy or paradoxically quiet breath sounds.

15 BRONCHIAL BREATH SOUNDS

1. When lung tissue between the central airways and the chest wall becomes airless because of conditions that increase lung density, transmission of breath sounds from large airways is enhanced.

2. Breath sounds are enhanced over an area of increased lung density because little high-frequency sound is lost through attenuation or filtration. As lung tissue density increases the impedance between the fluid-filled lung tissue and the pleurae and chest wall, these three media become well matched. This match decreases the normal filtering of high-frequency sounds. Consequently, the breath sounds are transmitted more readily to the chest wall surface and are louder and more tubular than normal breath sounds heard over the same chest wall area.

3. Conditions associated with bronchial breath sounds include consolidation, atelectasis, and fibrosis.

4. Clinical findings vary, depending on the location of the consolidated area and the causative agent. However

when classic consolidation is present, decreased chest wall movement and a dull percussion note are apparent over the affected area.

5. Bronchial breathing is heard over a dense, airless upper lobe, even without a patent bronchus, because the upper lobe surfaces are in direct contact with the trachea and loud tracheal breath sounds are transmitted directly to the dense, airless upper-lobe tissues. In contrast, bronchial breath sounds are heard over a dense, airless lower lobe only when the bronchi are patent because sound isn't transmitted directly to the airless lower-lobe tissues.

6. Bronchial breath sounds heard over a consolidated area are high pitched and have the typical hollow, or tubular, quality of normal central airway breath sounds. They remain audible during both expiration and inspiration, but the expiratory sounds are longer and louder when the patient is sitting up. The I:E ratio is 1:2. These sounds may be auscultated using either the bell or diaphragm of the stethoscope.

7. Atelectasis, incomplete expansion of a lung area, is thought to result from uncleared secretions that become thick, occluding the airway, or from hypoventilation.

8. Bronchial breath sounds heard over an atelectatic area are high pitched and have the typical hollow, or tubular, quality of normal central airway breath sounds. They're audible throughout inspiration and expiration. When the patient is in the supine position, the inspiratory and expiratory sounds are equal in duration and intensity. The I:E ratio is 1:1. These sounds can be heard equally well using either the bell or diaphragm of the stethoscope.

9. Bronchial breath sounds heard over a fibrotic area have the typical hollow, or tubular, quality of normal central airway breath sounds. They're audible throughout inspiration and expiration. When the patient is in the upright position, the breath sounds become progressively louder during inspiration and become both louder and longer during expiration. The I:E ratio is 1:1 to 1:2. These bronchial breath sounds are high pitched and are heard equally well using either the diaphragm or bell of the stethoscope.

16 ABNORMAL VOICE SOUNDS

1. Voice sounds arc produced by vibrations of the vocal cords as air from the lungs passes over them. The resonance of the mouth, nasopharynx, and paranasal sinuses helps to amplify these sounds.

2. Transmitted voice sounds are normally heard as a low-pitched, unintelligible mumble over healthy lung and pleural surfaces because most of the vowels are filtered out.

3. Voice sounds are heard distinctly over consolidated or atelectatic lung tissue areas because less filtering takes place, enhancing transmission.

4. Bronchophony is the clear, distinct, intelligible voice sound heard over dense, airless lung tissue. Dense, airless lung tissue transmits high-frequency vowel sounds more easily because of impedance matching. Also, consolidation increases vocal resonance, which allows the clear transmission of voice sounds to the chest wall.

5. Bronchophony is heard over dense, airless upper lobes because the upper-lobe surfaces are in direct contact with the trachea, leading to direct transmission of tracheal breath sounds. In contrast, bronchophony is heard over dense, airless lower lobes only when the bronchi are patent because a direct path of sound transmission to the lower lobes doesn't exist.

6. Bronchophony can be heard anywhere over the anterior, lateral, and posterior chest wall surfaces. It's commonly heard over dense, airless lung tissue in the upper lobes, such as a consolidated area.

7. The patient is asked to repeat the words "ninety-nine" several times. Over healthy lung tissue, the words are unintelligible; however, over a consolidated area, the high-frequency sounds are easily understood as words.

8. Whispered pectoriloquy is the clear, distinct, intelligible whispered voice sound heard over airless, consolidated, or atelectatic lung tissue. In a healthy person, normal lung tissue filters the high frequencies of whispered vowel sounds, making them unintelligible during auscultation. However, in a patient with consolidation or atelectasis, these same whispered vowel sounds are transmitted to the chest wall surface without much filtering of high frequencies and can be heard clearly using the stethoscope.

9. Whispered pectoriloquy can be heard anywhere over the anterior or posterior chest wall surfaces over dense, airless lung tissue, typically over an area of consolidation or atelectasis.

10. Egophony, a voice sound that has a nasal or bleating quality when heard over the chest wall, is produced over consolidated and atelectatic areas because impedance matching enhances breath sound transmission in such areas. It's also detectable over the upper border of large pleural effusions.

11. Egophony can be heard anywhere over the anterior, posterior, or lateral chest wall surfaces over a consolidated or atelectatic area.

12. The patient is asked to repeat the letter "E" several times. Over healthy lung tissue, the letter sounds as it normally does; over the consolidated area, it sounds like "A-aay" and has a high-pitched, nasal quality.

13. Bronchial breath sounds, bronchophony, whispered pectoriloquy, and egophony are typical in patients with consolidation.

17 ABSENT AND DIMINISHED BREATH SOUNDS

1. Breath sounds are diminished or eliminated by conditions that limit airflow into lung segments. Slow inspiratory airflow rates decrease breath sound amplitude because less air movement occurs, resulting in less turbulent airflow. Diminished or absent breath sounds can also occur when breath sounds are reflected at the visceral and parietal pleurae because of an impedance mismatch.

2. Conditions associated with absent or diminished breath sounds include shallow breathing, diaphragmatic paralysis, severe airway obstruction, pneumothorax, pleural effusion, hyperinflated lungs, and obesity. The use of positive end-expiratory pressure during assisted ventilation is also associated with diminished breath sounds.

3. During normal breathing in the upright position, a certain amount of air flows through the airways during inspiration and expiration; the distribution of ventilation is greater in dependent lung regions because more respiratory movement occurs. During shallow breathing, less respiratory movement occurs; consequently, less air flows through the airways during inspiration and

expiration. Because of this reduced airflow, turbulence is decreased and breath sounds are diminished.

4. When the diaphragm becomes paralyzed, as in a patient with a mediastinal tumor, it no longer participates in normal breathing. The internal and external intercostal muscles, which also have a role in normal breathing, must take over the work of breathing. With only the chest wall muscles initiating the respiratory cycle, ventilation of the lung bases may be limited, resulting in diminished breath sounds.

5. If a lobar or segmental bronchus becomes obstructed, as in a patient who has aspirated a foreign object, airflow ceases distal to the obstruction; therefore, breath sounds will be absent over the area distal to the obstruction. If a mainstem bronchus is obstructed, breath sounds will be absent throughout the affected lung.

6. Breath sounds heard over a pneumothorax are significantly diminished or absent because of acoustical mismatching of the air-filled lung and the collection of air in the pleural space.

7. Diminished or absent breath sounds resulting from a pneumothorax can be heard anywhere over the anterior, posterior, and lateral chest wall surfaces, depending on the location and size of the pneumothorax. Normal breath sounds are heard on the contralateral side over healthy lung tissue, and breath sounds are absent over the pneumothorax. If diminished breath sounds are heard, they have a low pitch that's heard using either the bell or diaphragm of the stethoscope.

8. In pleural effusion, fluid accumulates in the pleural space, impairing the transmission of normal breath sounds.

9. A large pleural effusion compresses adjacent lung tissue, causing atelectasis. The percussion tone is dull. Egophony, bronchophony, and whispered pectoriloquy caused by the atelectasis may be audible at the upper border of the pleural effusion. Occasionally, loud bronchial breath sounds may have sufficient intensity to be transmitted through a small pleural effusion.

10. Diminished or absent breath sounds resulting from a pleural effusion may be heard anywhere over the anterior, posterior, or lateral chest wall surfaces, depending on the location of the pleural effusion. Normal breath sounds are heard on the contralateral side over healthy lung tissue. The diminished breath sounds have a low

pitch that can be heard using either the bell or diaphragm of the stethoscope.

11. Premature dynamic compression of the large central airways and the loss of elastic tension in the lung tissue limit airflow during expiration. Increased intrinsic airway resistance may also be present. Increased amounts of air trapped in the lungs change the acoustical qualities of the sound transmitted, creating a mismatch between the pleurae and chest wall and the hyperinflated lung tissue. Consequently, breath sounds are diminished.

12. Diminished or absent breath sounds associated with chronic obstructive pulmonary disease are heard over the anterior, posterior, and lateral chest wall surfaces. If the breath sounds are audible, they're soft in intensity and have a low pitch that's heard best using the diaphragm of the stethoscope. The sounds are heard throughout inspiration and expiration.

13. Breath sounds are normally attenuated and filtered at the pleural surface. In an obese person, a thickened chest wall increases the distance between lung tissue and the chest wall surface. This results in additional filtering of the breath sounds as they are transmitted from the pleurae to the chest wall surface.

14. Adding positive end-expiratory pressure (PEEP) during mechanical ventilation of an intubated patient can diminish breath sounds in the following way: PEEP increases functional residual capacity (FRC), the amount of air remaining in the airways at the end of normal expiration. Increasing FRC hyperinflates the lungs. Increased amounts of air in the small airways and alveoli change the acoustical qualities of the sound transmitted through the chest wall, creating a mismatch between the pleurae and chest wall, and the hyperinflated lung.

18 CLASSIFYING ADVENTITIOUS SOUNDS

1. In 1977, the American Thoracic Society adopted a system of classifying adventitious sounds based on acoustical qualities, timing, and frequency waveforms.

2. Adventitious sounds are classified as crackles and wheezes.

3. Coarse crackles are discontinuous, explosive sounds that are loud and low pitched.

4. Fine crackles are discontinuous, explosive sounds that are shorter in duration, higher in pitch, and less intense than coarse crackles.

5. Wheezes are continuous sounds that are high pitched and have a hissing or coughing sound. They commonly have a musical quality.

6. Low-pitched wheezes are continuous sounds that are low pitched and commonly resemble snoring.

19 ADVENTITIOUS SOUNDS: CRACKLES

1. Crackles are short, explosive or popping sounds that are described according to their pitch, timing, and location.

2. Two different mechanisms are thought to generate crackles: air bubbling through secretions and airways opening suddenly and explosively.

3. Air bubbling through secretions in the airways is widely accepted as one way crackles are produced. According to Forgacs, this explanation is probably true when the trachea and mainstem bronchi are full of sputum, as commonly occurs with severe pulmonary edema and chronic bronchitis.

4. The sudden opening of multiple collapsed peripheral airways and the associated explosive changes in air pressures are thought to produce the crackles heard over the lung bases in a healthy person who inhales deeply after a maximum exhalation.

5. The timing of crackles is described as early inspiration, midinspiration, or late inspiration or early expiration, midexpiration, or late expiration. Pitch is usually described as high or low. Intensity varies and is described as loud or soft. Density is described as profuse or scanty. Duration, indicating the length of time that the crackles can be heard during inspiration or expiration, varies.

6. Conditions associated with late inspiratory crackles include atelectasis, resolving lobar pneumonia, interstitial fibrosis, and left-sided heart failure.

7. Crackles associated with atelectasis are produced by the sudden opening of collapsed small airways and adjoining alveoli.

8. The crackles associated with atelectasis begin late in inspiration and become more profuse toward the end of inspiration. They vary in intensity but are high

pitched. Because crackles associated with atelectasis are poorly transmitted to the chest wall surface, their intensity and density change when the stethoscope is moved only a short distance. They aren't audible at the mouth. The patient's position also affects these crackles. For example, in an immobile patient, profuse crackles are heard in the dependent lung regions, but crackles are absent or scanty in nondependent lung regions. Also, crackles associated with atelectasis may clear somewhat with coughing.

9. In patients with resolving lobar pneumonia, crackles can be auscultated over lung areas where many alveoli are still filled with exudate, but the surrounding alveoli are aerated and have higher-than-normal ventilation. A large increase in air pressure gradients in the airways reaching these unaerated alveoli generates crackles as the airways are snapped open during late inspiration.

10. Crackles associated with lobar pneumonia begin late in inspiration and become more profuse toward the end of inspiration. These crackles are typically high pitched.

11. Diffuse interstitial fibrosis impairs or destroys alveoli by filling them with abnormal cells or by scarring the lung tissue. Unaffected alveoli are usually hyperaerated. The lungs become stiff and more difficult to inflate, and airflow volumes usually decrease. Theoretically, the delayed opening of small airways during inspiration causes alveolar pressures in the affected lung tissue to fall more significantly than in healthy alveoli. This fall in alveolar pressures leads to an increased pressure gradient that generates repetitive late inspiratory crackles.

12. As interstitial fibrosis worsens, crackles are heard bilaterally over the posterior lung bases and spread upward toward the apices. In later stages of the disease, crackles aren't affected by position changes and may be heard throughout inspiration.

13. Crackles associated with interstitial fibrosis have a fine intensity and a short, discontinuous duration; they're heard during late inspiration. They have a high pitch that's heard best using the diaphragm of the stethoscope.

14. Left-sided heart failure leads to fluid accumulation in the lung interstitium. Crackles are produced by the rapid equalization of pressures associated with the de-

layed opening of airways narrowed by pulmonary edema.

15. Crackles associated with early left-sided heart failure and pulmonary edema are heard bilaterally over the posterior lung bases. However, as pulmonary edema worsens, the crackles become more profuse and may be heard throughout the chest during late inspiration. Their intensity is fine in the early stages of left-sided heart failure but changes to a loud rattling sound as the patient's condition worsens. Their duration varies with the degree of left-sided heart failure. They have a low pitch that's heard best using either the bell or diaphragm of the stethoscope.

16. Coarse crackles are thought to be produced by the intermittent closure of large bronchi and the corresponding flow of a bolus of air through the obstructed area.

17. Early inspiratory crackles are associated with excessive mucus production in patients with chronic bronchitis. The proliferation and hypertrophy of mucous glands within the airways are caused by chronic exposure to airway irritants, such as cigarette smoke and air pollution.

18. In patients with chronic bronchitis, crackles are heard early in inspiration over all chest wall surfaces and at the mouth. They are scanty and low pitched and are not affected by the patient's position.

19. Bronchiectasis, irreversible dilation of the bronchi in selected lung segments, is characterized by chronic, copious production of yellow or green sputum and by fibrotic or atelectatic lung tissue surrounding the affected airways.

20. In patients with bronchiectasis, crackles tend to be profuse, low pitched, heard during early inspiration or midinspiration, and coarser than those associated with chronic bronchitis. Coughing may decrease the number of crackles heard, but position changes don't affect them.

21. When the visceral and parietal pleural surfaces are damaged by fibrin deposits or inflammatory or neoplastic cells, they lose their ability to glide silently over each other during breathing, and their movements become jerky and periodically delayed. This motion produces loud, grating crackles known as pleural crackles or pleural friction rub.

22. Pleural crackles are heard on the chest wall over the affected area. Their intensity is loud, and they have a coarse, grating quality and a low pitch. Their duration is discontinuous. They may be heard during inspiration only or during both the inspiratory and expiratory phases of the respiratory cycle.

20 ADVENTITIOUS SOUNDS: WHEEZES

1. Wheezes are musical sounds generated by air passing through a bronchus so narrowed as to be almost closed. The bronchus walls oscillate between closed and barely open positions; these oscillations generate audible sounds.

2. Wheezes are high-pitched, continuous sounds with frequencies of 200 Hz or greater and a duration of 250 msec or more. Their duration is long enough to carry an audible pitch similar to a musical tone. Wheezes are described by their timing within the respiratory cycle: They may be heard during inspiration or expiration or continuously throughout the respiratory cycle. They can be further described as localized or diffuse, episodic or chronic.

3. The pitch is determined by the emitted sound's frequency, which can vary widely over a five-octave range. These differences in frequency are attributed to the airway size and elasticity and to airflow rates through the narrowed bronchus.

4. Wheezes are transmitted better through airways than through lung tissue, which absorbs high-frequency sounds.

5. Conditions associated with wheezes include bronchospasm, airway thickening caused by mucosal swelling or muscle hypertrophy, inhalation of a foreign object, tumor, secretions, or dynamic airway compression.

6. Polyphonic wheezes are believed to be caused by the dynamic compression of large airways during expiration. In healthy individuals, polyphonic wheezes can sometimes be auscultated during a maximal forced expiration when the dynamic compression occurs simultaneously in all airways.

7. In patients with widespread air-flow obstruction, elastic recoil properties, peripheral airway resistance, and airway mechanics are altered throughout both lungs.

Together these abnormal changes affect the timing of dynamic airway compression within the bronchi. This change in airway compression alters normal sound production.

8. Expiratory polyphonic wheezes are usually heard over the anterior, posterior, and lateral chest wall surfaces during expiration. Their intensity is usually described as loud and musical, and their duration is continuous. They have a high pitch that's heard best using the diaphragm of the stethoscope.

9. Fixed monophonic wheezes, which have a constant pitch and a single musical tone, are generated by the oscillations of a large, partially obstructed bronchus. The obstruction may be caused by a tumor, a foreign body, bronchial stenosis, or an intrabronchial granuloma.

10. Fixed monophonic wheezes are heard over the anterior, posterior, and right lateral chest wall surfaces. These wheezes are usually loud, and their duration is continuous. They may be heard during inspiration or expiration or throughout the respiratory cycle. They have a low pitch that's heard best using the diaphragm of the stethoscope.

11. Sequential inspiratory wheezes are sometimes heard over the lung bases in patients with interstitial fibrosis, asbestosis, or fibrosing alveolitis.

12. These monophonic wheezes are generated by airways that open late in inspiration in unaerated lung regions. The rapid inflow of air precipitates airway wall vibrations, generating sequential inspiratory wheezes.

13. Sequential inspiratory wheezes are usually heard over the lateral and posterior lung bases. They have a loud intensity and a continuous duration. They occur throughout inspiration but are more predominant in late inspiration. They have a high pitch that's heard best using the diaphragm of the stethoscope.

14. Airways narrowed by bronchospasm or mucosal swelling produce single or multiple monophonic wheezes that may occur during inspiration or expiration or throughout the respiratory cycle.

15. Random monophonic wheezes are typically heard in patients with severe status asthmaticus.

16. The progressive airway obstruction that occurs in patients with status asthmaticus can cause a dynamic pattern of wheezing. First, monophonic wheezes are

heard only during expiration; then they are heard throughout the respiratory cycle. As status asthmaticus becomes more severe, all wheezes heard over the chest wall surfaces disappear because a combination of air trapping and severe airway narrowing causes the site of dynamic airway compression to move toward the lung periphery. This phenomenon is called a silent chest.

17. Random monophonic wheezes are usually heard over the anterior, posterior, and lateral chest wall surfaces. They're usually loud, and their duration is continuous. These wheezes occur throughout the respiratory cycle, and expiration is commonly prolonged. They have a high pitch that's heard best using the diaphragm of the stethoscope.

18. Stridor is a loud musical sound that's heard at a distance from the patient, usually without a stethoscope.

19. Stridor is caused by laryngeal spasm and mucosal swelling, which contract the vocal cords and narrow the airway.

20. Severe stridor is heard without a stethoscope; however, in a patient with a less pronounced laryngeal spasm, stridor may be heard by auscultating over the larynx. Stridor is loud and has a continuous duration. It may be heard during inspiration or throughout the respiratory cycle. Its high pitch resembles a crowing sound.

Selected readings

ACLS Provider Manual. Dallas: American Heart Association, 2003.

Anatomy & Physiology Made Incredibly Easy, 2nd ed. Philadelphia: Lippincott Williams & Wilkins, 2005.

Assessment Made Incredibly Easy, 3rd ed. Philadelphia: Lippincott Williams & Wilkins, 2005.

Critical Care Nursing Made Incredibly Easy. Philadelphia: Lippincott Williams & Wilkins, 2004.

Finesilver, C. "Pulmonary Assessment: What You Need to Know," *Progress in Cardiovascular Nursing* 18(2):83-92, Spring 2003.

Hockenberry, M.J. *Wong's Nursing Care of Infants and Children,* 7th ed. St. Louis: Mosby–Year Book, Inc., 2003.

Jarvis, C. *Physical Examination & Health Assessment,* 4th ed. St. Louis: W.B. Saunders Co., 2004.

Karnath, B. "Cardiac Murmurs: Diagnostic Clues You Can Pick Up in a Heartbeat," *Consultant* 44(3):425-27, 431-32, March 2004.

LeMone, P., and Burke, K. *Medical-Surgical Nursing: Critical Thinking in Client Care,* 3rd ed. Upper Saddle River, N.J.: Prentice Hall Health, 2004.

London, M.L., et al. *Maternal-Newborn and Child Nursing: Family Centered Care.* Upper Saddle River, N.J.: Prentice Hall Health, 2003.

Mehta, M. "Assessing Respiratory Status," *Nursing2003* 33(2):54-56, February 2003.

Miller, C. *Nursing for Wellness in Older Adults: Theory and Practice,* 4th ed. Philadelphia: Lippincott Williams & Wilkins, 2004.

Price, S.A., and Wilson, L.M., eds. *Pathophysiology: Clinical Concepts of Disease Processes,* 6th ed. St. Louis: Mosby–Year Book, Inc., 2003.

Porth, C.M., and Kunert, M.P. *Pathophysiology: Concepts of Altered Health States,* 6th ed. Philadelphia: Lippincott Williams & Wilkins, 2002.

Smeltzer, S.C., and Bare, B.G. *Brunner and Suddarth's Textbook of Medical-Surgical Nursing,* 10th ed. Philadelphia: Lippincott Williams & Wilkins, 2004.

Woodrow, P. "Assessing Respiratory Function in Older People," *Nursing Older People* 14(3):27-28, May 2002.

Glossary

A₂ *abbr* Aortic component of S_2.

Accessory muscles of respiration Muscles involved in labored, or forceful, breathing. They include the sternocleidomastoid, scalene, trapezius, rhomboid, and abdominal muscles.

Acoustical mismatch Phenomenon in which sounds transmitted through two different media, such as air and tissue, are diminished in intensity or blocked.

Acute pulmonary hypertension Sudden increased pressure within the pulmonary circulation (above 30 mm Hg systolic and 12 mm Hg diastolic).

Adult respiratory distress syndrome Disease with numerous causes that's characterized by interstitial and alveolar edema and progressive hypoxemia; also known as *shock lung* or *posttraumatic lung.*

Adventitious sound An acquired (usually) abnormal sound superimposed over normal or abnormal breath sounds.

AES *abbr* Aortic ejection sound.

Alveolar-capillary membrane Cell membrane across which oxygen and carbon dioxide must diffuse for respiration to occur.

Alveolar duct The narrowed end portion of the bronchioles that terminate in clusters of alveolar sacs.

Alveolus One of the millions of small, saclike lung structures where oxygen–carbon dioxide exchange occurs.

Amplitude Magnitude or intensity of a sound or pulsation.

Anatomic dead space Air that remains in the conducting airways during each breath. This air isn't involved in oxygen–carbon dioxide exchange.

Antigen Substance or material foreign to the body that causes a reaction leading to the formation of antibodies.

Aortic ejection sound Opening sound of a stenotic aortic valve. It follows S_1 early in ventricular systole and appears just after the QRS complex on the ECG.

Aortic regurgitation Abnormal condition of turbulent backward blood flow through the aortic valve into the left ventricle during diastole.

Aortic stenosis Narrowing or constriction of the aortic valve or of the aorta itself.

Aortic valve Membranous folds that prevent blood reflux from the aorta into the left ventricle.

Aortic valvular stenosis Constriction of or damage to the aortic valve that restricts forward blood flow from the left ventricle to the aorta during systole.

ARDS *abbr* Acute respiratory distress syndrome.

Arteriovenous shunt Direct passage of blood from arteries to veins, bypassing the capillary bed. This can refer to a physiologic response of the body or to an abnormal condition sometimes caused by trauma or surgery.

Asbestosis Lung disease characterized by pulmonary inflammation and fibrosis; caused by prolonged asbestos exposure.

Ascending aortic aneurysm Dilation of the thoracic portion of the aorta.

Aspiration Usually refers to the breathing in of foreign materials.

Asthma Disease characterized by bronchoconstriction, bronchospasm, mucosal edema, and excess mucus production, which lead to obstructed airflow, wheezing, and shortness of breath.

Atelectasis Incomplete expansion of the lung tissue, usually caused by pressure from exudate, fluid, tumor, or an obstructed airway; may involve a lung segment or an entire lobe.

Atrial septal defect Imperfection, failure to develop fully, or absence of the dividing wall (septum) between the heart's atria.

Atrioventricular node Small mass of specialized cardiac tissue, located in the lower portion of the right atrium near the septum, that transmits electrical impulses from the sinoatrial node to the bundle of His.

Atrioventricular valves Valves between the atria and the ventricles — specifically, the tricuspid valve of the heart's right side and the mitral valve of the heart's left side.

Attenuation Decrease in the intensity or loudness of a breath sound.

Auscultation Act of listening to sounds made by the body; usually performed with a stethoscope.

Austin Flint murmur Diastolic heart sound generated by turbulent blood flow across the mitral valve; caused by aortic regurgitation, which closes the mitral leaflets.

AV *abbr* Atrioventricular.

Axillary fold One of two anatomic landmarks anterior and posterior to the armpit, formed by the normal contour of the skin over the pectoralis major muscle anteriorly and the latissimus dorsi muscle posteriorly.

Ball-in-cage valve Prosthetic heart valve characterized by a caged ball that moves with ventricular pressure to open and close a valvular opening.

Bell Cup-shaped portion of the stethoscope that's best suited for listening to low-pitched sounds.

Bifurcate To divide into two branches.

Bileaflet valve (1) Prosthetic heart valve characterized by two small wings that control blood flow; (2) mitral valve.

Binaural headpiece Stethoscope headpiece that supplies sounds to both ears simultaneously.

Blowing Term used to describe a continuous murmur sound, like the sound of air passing through pursed lips.

Booming Term used to describe a deep, resonant heart sound that's sudden and percussive.

Bronchial circulation Oxygenated blood that arises from the aorta or subclavicular artery and supplies the tracheobronchial tree with oxygen and nutrients.

Bronchial gland One of the glands secreting mucus and serous liquid in the tracheobronchial tree. These glands are most numerous in the medium-sized bronchi and are the main source of bronchial secretions.

Bronchiectasis Irreversible dilation of the bronchi characterized by chronic cough, sputum production, fibrosis, and atelectatic lung tissue surrounding the affected airways.

Bronchophony Voice sound with increased tone or clarity in vocal resonance auscultated over the chest wall.

Bronchospasm Smooth-muscle contraction within the airway walls, which leads to airway narrowing and reduced airflow and may be accompanied by coughing and wheezing.

Bronchovesicular breath sound Normal breath sound auscultated between the mainstem bronchi and lung periphery; also known as a *transitional breath sound.*

Bundle of His Small band of specialized cardiac muscle fibers located in the intraventricular septum that relay the atrial electrical impulses to the ventricles.

Capacitance Heart's or vasculature's ability to receive or admit blood volume and the decreased resistance that accompanies it.

Capillary hydrostatic pressure Fluid pressure within the capillary system, which, when elevated, leads to fluid extravasation out of the capillary system into the interstitium.

Cardiac cycle Sequence of events (systole and diastole) that enables the heart to receive and pump blood.

Cardiac output Quantity of oxygenated blood pumped through the body by the heart, usually expressed in liters per minute. Is computed as stroke volume $\times$ heart rate.

Cardiac tamponade Heart compression caused by effusion or collection of fluid in the pericardium, resulting in decreased cardiac output.

Cervical venous hum murmur Murmur caused by rapid downward blood flow through the jugular veins in the lower part of the neck.

Chordae tendineae Tendinous cords that connect each cusp of the two atrioventricular valves to appropriate papillary muscles in the ventricles.

Chronic bronchitis Disease characterized by chronic cough and sputum production that persists over a long period of time.

Chronic obstructive pulmonary disease Chronic lung disease characterized by obstructed bronchial airflow or exhalation. Chronic bronchitis, emphysema, and asthma are types of obstructive lung disease.

Cilia Motile, whiplike extensions from cell surfaces. Ciliated columnar epithelial cells line the walls of the tracheobronchial tree.

Click Short heart sound; also, another term for a midsystolic sound that occurs when a prolapsed mitral valve's leaflet and chordae tendineae tense.

Click-murmur syndrome Condition in which a click is followed by a murmur, as in mitral valve prolapse syndrome.

Closing click Heart sound generated by valve closure. Closing clicks are heard with all prosthetic valves, regardless of their type or position.

Coarctation of the aorta Localized aortic malformation characterized by deformity of the aortic media and causing severe narrowing of the aortic lumen.

Compliance Tissue's or organ's ability to yield to pressure without disruption; commonly used to describe the distensibility of an air- or fluid-filled organ, such as the heart or lungs.

Conchae Shell-shaped turbinate bones in the nasal cavity that stimulate turbulent airflow within the nasal passages. The dense mucus and capillary beds that line the turbinates' surface encourage warming and humidification of inspired air.

Conducting airway One of the airways beginning at the nose and ending at the terminal bronchioles. These airways are responsible for transporting air during breathing but are not involved in oxygen–carbon dioxide exchange.

Conduction system Specialized cardiac cells and fibers that initiate and/or relay the electrical impulses, stimulating the heart muscle to contract.

Configuration Shape of a murmur's sound as recorded on a phonocardiogram; one of the characteristics used to describe murmurs.

Congestion Abnormal accumulation of fluid or blood in an organ or organ part.

Consolidation Inflammatory solidification of lung tissue.

Constrictive pericarditis Inflammation of the pericardium that leads to thickening and possible calcification of the pericardial sac, resulting in impaired diastolic filling, inflow stasis, or a constrictive effect.

Continuous murmur Murmur that begins in systole and persists, without interruption, through S_2 into diastole.

Contractile cell One of the myocardial cells that contract to start systole. These cells depolarize and repolarize by means of ion flow.

Contralateral Referring to the opposite side or opposing symmetrical structure.

COPD *abbr* Chronic obstructive pulmonary disease.

Cor pulmonale Heart disease caused by pulmonary hypertension secondary to disease of the lung or its blood vessels, resulting in hypertrophy of the right ventricle.

Crackle An adventitious lung sound characterized by a short, explosive or popping sound usually heard during inspiration; described as *coarse* (loud and low in pitch) or *fine* (less intense and high in pitch); formerly known as a *rale*.

Crepitation Crackling sound that resembles the sound made by rubbing hair between two fingers.

Crescendo Term used to describe the configuration of a murmur that increases in intensity.

Crescendo-decrescendo Term used to describe the configuration of a murmur that rises in intensity and then fades away.

Cusp One of the triangular segments of a cardiac valve or one of the semilunar segments of the aortic or pulmonary valve.

Dampened Diminished sound intensity or amplitude; term used to describe sounds.

Decrescendo Term used to describe the configuration of a murmur that decreases in intensity.

Dependent lung regions Lowest lung regions (the lung bases when the patient is upright).

Depolarization Movement of sodium ions into a contractile cell, creating a positive charge inside the cell.

Diamond-shaped murmur Crescendo-decrescendo murmur.

Diaphragm Primary muscle of respiration, which separates the thoracic and abdominal cavities; also the part of the stethoscope used to auscultate for high-pitched sounds.

Diastole Expansion of the ventricles occurring in the interval between S_2 and S_1. The heart muscle, relaxing, fills with blood during this part of the cardiac cycle.

Diffuse Widely distributed; not localized.

Dilation of pulmonic valve ring Expansion of a valve aperture. A dilated pulmonic valve can contribute to a Graham Steell murmur.

Distribution of ventilation Movement or circulation of air to specific lung regions during breathing.

Dull percussion note Deadened, or nonresonant, sound heard when a solid organ or dense body part is percussed.

Duration Length of time a heart or breath sound is heard; one of the six characteristics used to describe heart sounds.

Dynamic airway compression Narrowing of the airways during expiration caused by properties of intrapleural pressure, radial traction exerted by lung parenchyma, and the loss of elastic recoil within the lung.

Dynamic obstruction Blocked outflow from one of the heart's chambers that can be demonstrated only during myocardial contraction or systole.

Dyspnea Difficult or labored breathing.

Early diastolic aortic regurgitation Blood backflow from the aorta that occurs early in diastole, when the ventricle is resting and filling with oxygenated blood.

ECG *abbr* Electrocardiogram; also written as EKG.

Ectopic beats Arrhythmic heartbeats arising from places other than the heart's normal pacemaker, the sinoatrial node.

Edema Excessive accumulation of fluid in intercellular tissue spaces of the body.

Egophony Voice sound that has a nasal or bleating quality when auscultated over the chest wall; "ee" is heard as "ay;" an "A" to "E" change.

Ejection sound Sound caused by the opening of a stenotic aortic or pulmonic valve, usually occurring early in systole.

Ejection velocity Measure of the speed of blood flow.

Elastic recoil Spontaneous contraction of lung parenchyma that occurs during expiration and that helps move air out of the lungs.

Elastic tension Support and traction exerted on the airways because of the natural elastic recoil properties of the surrounding lung parenchyma.

Epicardial pacing Regulation of the rate of heart muscle contraction by an artificial cardiac pacemaker stimulating the heart through electrical leads attached to the heart surface.

Epiglottis Small elastic cartilage attached at the larynx that covers the opening to the trachea during swallowing.

Equal pressure point Point in the airways where intrapulmonary and intrapleural pressures are equal.

ES *abbr* Ejection sound.

Exudate Inflammatory fluid leaked from body cells or tissues.

Fibrosis Abnormal formation of fibrous connective tissue that usually occurs as a reparative or reactive process within an organ or tissue but ultimately replaces healthy, functional tissue.

5LMCL *abbr* Fifth intercostal space, left midclavicular line (mitral area).

5LSB *abbr* Fifth intercostal space, left sternal border (tricuspid area).

FRC *abbr* Functional residual capacity.

Frequency Pitch of a breath sound measured in hertz.

Functional murmur Benign murmur that doesn't impair heart function.

Gallop rhythm Triple rhythms; the S_1, S_2, S_3 sequence; the S_4, S_1, S_2 sequence; or both. Sounds like a horse's gallop.

Gas exchange surface Alveolar capillary surface that's actively involved in diffusing oxygen and carbon dioxide.

Graham Steell murmur Pulmonic regurgitation murmur caused by pulmonic hypertension and pulmonic valve ring dilation.

Heart failure Clinical syndrome caused by left-sided or right-sided heart dysfunction. Left-sided heart failure results in pulmonary edema and breathlessness. Right-sided heart failure results in liver congestion, increased venous pressure, and peripheral edema.

Hemidiaphragm One half of the diaphragm (either the right or left side).

High-output condition Physiologic state that causes or results in increased cardiac output.

Hila Medial lung aspects where the bronchi, nerves, and vessels enter and leave.

Holodiastolic Pertaining to the entire diastole; used to describe a murmur that persists throughout diastole.

Holosystolic Pertaining to the entire systole; used to describe a murmur that persists throughout systole.

Holosystolic mitral regurgitation Blood backflow that's caused by an incompetent mitral valve and that can be heard throughout systole.

Hyperinflation Overinflation of the lung that occurs with air trapping in obstructive lung diseases, such as emphysema.

Hypertrophic cardiomyopathy Narrowing or constriction of the left ventricle's subaortic region caused by tissue enlargement in that area; also known as *idiopathic hypertrophic subaortic stenosis*. This condition can be a cause of subvalvular aortic stenosis.

Hypertrophy Enlargement or overgrowth of an entire organ or of part of an organ, caused by an increase in the size of its constituent cells.

Hypoxemia Abnormally low oxygen tension in arterial blood.

Hz *abbr* Hertz.

Idiopathic Of unknown cause.

I:E ratio *abbr* Inspiratory-expiratory ratio.

Impedance matching Similar acoustical characteristics of two organs or types of tissue that allow effective sound transmission.

Incompetent valve Heart valve that cannot perform its functions because of congenital defects, disease, or trauma.

Infective endocarditis Inflammation of the endocardium caused by infection with microorganisms (bacteria or fungi); primarily affects the heart valves.

Inferior vena cava Venous trunk for the lower extremities and the pelvic and abdominal viscera.

Inspiratory-expiratory ratio Numerical expression of the duration of inspiration in relation to the duration of expiration.

Intensity Degree of loudness; one of six characteristics used to describe heart sounds.

Intercostal muscle One of the muscles found between the ribs. Internal and external intercostal muscles help stabilize and expand or lower the rib cage with ventilation.

Internodal pathway One of the fibers connecting the small masses of tissue that transmit the electrical impulses setting the heart's rate and rhythm.

Interstitial fibrosis Abnormal formation of fibrous tissue that occurs as a reparative or reactive process within the alveolar septa and interstitial areas of the lungs.

Interstitium Small gap in an organ or a tissue; in lung parenchyma, the space between the alveolar and capillary membranes.

Interventricular septum Dividing wall between the heart's ventricles.

Intrapleural pressure Relative pressure that occurs between the pleurae. Negative pressure occurs during inspiration; positive pressure occurs during expiration.

Intrapulmonary pressure Pressure within the lung. Negative pressure causes air to flow inward; positive pressure causes air to move outward.

Isovolumic contraction Initial period in early systole when the atrioventricular and semilunar valves are closed and intraventricular pressures rise but blood has not yet been ejected from the ventricles.

Isovolumic relaxation Brief period in early diastole when the atrioventricular and semilunar valves are closed, just before the atrioventricular valves open and passive filling of the ventricles begins.

Laminar airflow Orderly, linear airflow.

Larynx Cartilaginous organ located between the pharynx and the trachea that houses the vocal cords and allows for voice production.

LBBB *abbr* Left bundle-branch block.

Leaflet Structure resembling a small leaf, especially a heart valve's cusps.

Left atrial shunt Small amount of deoxygenated blood that returns to the left atrium; normally, this is venous return from bronchial circulation.

Left bundle-branch block Interrupted conduction through the fiber that activates the left ventricle, resulting in a prolonged or abnormal QRS complex.

Left lateral recumbent position Position that brings the heart's apex closer to the chest wall for auscultation. The patient lies on his left side, with his right knee and thigh drawn up to his chest.

Left mainstem bronchus One of two main branches extending from the trachea that supplies air to the left lung. It leaves the trachea at a sharper angle than the right mainstem bronchus and passes under the aortic arch before entering the lung.

Left-sided heart failure Inability of the left ventricle to pump blood adequately, causing decreased cardiac output, which results in pulmonary congestion and edema.

Left ventricular decompensation Failure of the left ventricle's myocardium to contract and maintain adequate cardiac output.

Left ventricular hypertrophy Enlargement of the left ventricle's myocardium, which can result in reduced or abnormal ventricular functioning.

Left ventricular outflow obstruction Stenosis or other blockage preventing flow of oxygenated blood from the left ventricle into the aorta.

Left ventricular pressure overload Increased pressure within the left ventricle that impairs its ability to function normally; commonly caused by excessively elevated systemic blood pressure.

Left ventricular volume overload Excessive volume within the left ventricle that causes it to become distended and impairs its ability to function normally.

Lobar Referring to or involving any lung lobe.

Location Site at which a breath or a heart sound can be auscultated.

Lower airway *See* Tracheobronchial tree.

Low-pitched wheeze Continuous, low-pitched sound that resembles snoring; previously classified as a sonorous bronchus or a sonorous rale.

M₁ *abbr* Mitral component of S_1.

Macrophage Large ameboid, mononuclear cell that acts as a defense mechanism against infection; one of three cell types lining the alveoli.

Mainstem bronchi breath sound Harsh, tubular (hollow) breath sound heard over a mainstem bronchus; also known as *rhonchi.*

Mast cell Connective tissue cell whose specific physiologic function remains unknown. The secretory mast cells on the surface of the large airways may be stimulated or affected by drugs, hormones, antigens, or other messenger cells to precipitate or control bronchoconstriction.

Mean airflow velocity Air flow rates occurring within an airway during the middle part of exhalation.

Mechanical ventilation Breathing that's assisted or controlled by a machine (a ventilator).

Mediastinal crunch Heart sound created by heart movement against air in the mediastinum; also known as *Hamman's sign.*

Mediastinum Tissues separating the two lungs, between the sternum and the vertebral column and from the thoracic inlet to the diaphragm. It contains the heart and its vessels, the trachea, esophagus, thymus, lymph nodes, and other organs and tissues.

Middiastolic aortic regurgitation Turbulent backward blood flow through the aorta into the left ventricle occurring midway between S_2 and S_1.

Middiastolic click Click heard midway between S_2 and S_1 in the cardiac cycle; usually associated with a stenotic or rigid atrioventricular valve.

Midsystolic ejection murmur Crisp, high-frequency sound resulting from a stenotic aortic or pulmonic valve opening in systole.

Mitral regurgitation Turbulent backward blood flow from the left ventricle into the left atrium caused by the mitral valve's inability to close completely.

Mitral stenosis Mitral valve narrowing or constriction obstructing flow of blood from the left atrium to the left ventricle.

Mitral valve Tissue folds that, when closed, prevent blood flow from the left ventricle to the left atrium; also called the *bicuspid valve.*

Mitral valve prolapse Mitral valve bulging from the proper position back toward the left atrium during systole. It can result in mitral regurgitation and may produce a late systolic murmur, possibly with a nonejection midsystolic click.

Monophonic Having one distinct musical sound or tone; used to describe selected wheezes.

MSC *abbr* Midsystolic click.

Mucus Serous, watery liquid secreted by bronchial glands and goblet cells within the airways.

Murmur Sound heard during auscultation that results from vibrations produced by blood moving within the heart and adjacent blood vessels; may be benign or abnormal.

Myocardial infarction Tissue death in the heart's muscle; usually caused by inadequate coronary artery perfusion.

Myocardium Heart wall, comprised of cardiac muscle tissue.

Nonejection click Click caused by a valve that isn't associated with movement of blood through or across the valve; commonly used to describe the click of mitral valve prolapse.

Normal breath sound Sound auscultated over chest wall areas of a healthy person.

Normal sinus rhythm Normal physiologic heart rhythm originating in the sinoatrial node; usually considered to be 60 to 100 beats/minute.

O_2-CO_2 exchange *abbr* Oxygen-carbon dioxide exchange.

Opening click Opening sound created by some prosthetic heart valves or by a diseased valve.

Opening snap Sound created by mitral valve leaflets that have become stenotic or abnormally narrowed but that are still somewhat mobile.

OS *abbr* Opening snap.

Oscillation Vibration, fluctuation.

P_2 *abbr* Pulmonic component of S_2.

Papillary muscle Muscular protrusion, or projection, in the ventricles that attaches to and regulates the atrioventricular valves by way of the chordae tendineae.

Parasympathetic nervous system Part of the involuntary nervous system that supplies innervation to the internal organs. The hormone mediator is acetylcholine. The vagus nerve is a parasympathetic nerve that innervates the airways and, when stimulated, causes smooth-muscle contraction, cough, and mucus discharge from the bronchial glands.

Parenchyma Functioning cells of an organ that distinguish or determine the primary organ function.

Patent ductus arteriosus Abnormal channel connecting the pulmonary artery to the descending aorta; results in arterial blood recirculation through the lungs.

PCG *abbr* Phonocardiogram.

PDA *abbr* Patent ductus arteriosus.

PEEP *abbr* Positive end-expiratory pressure.

Perfusion Blood flow to or through an organ or tissue supplied by the blood vessels.

Pericardial fluid Fluid in the space between the visceral and parietal layers of the pericardium.

Pericardial friction rub Characteristic high-pitched friction noise created by inflamed or dry pericardial surfaces rubbing together.

Pericardial knock S_3 associated with constrictive pericarditis.

Pericarditis Inflammation of the pericardium.

Pericardium Two-layer fibrous sac that surrounds the heart and the roots of the great vessels consisting of the visceral and parietal layers

Peripheral Toward the outer boundary or perimeter; not central.

Peripheral vascular resistance Resistance to the passage of blood through the small blood vessels, especially the arterioles, caused by friction between the blood and the blood vessel wall.

Pharynx Musculomembranous passage between the posterior nares and the larynx and esophagus; also referred to as the *throat.* It serves as a joint conduit for food and air.

Phonation Production of vocal sounds.

Phonocardiogram Graphic record of heart sounds and murmurs, including pulse tracings, produced by phonocardiography.

Physiologic S_3 S_3 that isn't associated with an abnormal condition; often found in young individuals.

Pitch A tone's vibration or frequency, measured in cycles per second as sound amplitude; subjectively described as high, medium, or low.

Plateau-shaped murmur Murmur in which intensity is the same, or flat, throughout its duration.

Pleurae Thin, serous membranes that surround the lungs (visceral pleura) and line the thoracic cavity's inner walls (parietal pleura).

Pleural crackle Loud, grating sound caused by inflamed or damaged pleurae; also called *pleural friction rub.*

Pleural effusion Abnormal accumulation of fluid between visceral and parietal pleurae.

Pleural friction rub Sound created by friction between the parietal and visceral pleurae surrounding the lungs; similar to a pericardial friction rub.

PMI *abbr* Point of maximum impulse.

Pneumonia Inflammation of the lung parenchyma.

Pneumothorax Accumulation of air within the pleural cavity.

Point of maximum impulse Chest wall site where the heartbeat can be seen or felt most strongly; typically located over the heart's apex.

Polyphonic Having multiple distinct musical sounds or tones; used to describe selected wheezes.

Porcine valve Prosthetic heart valve made from swine (pig) products.

Positive end-expiratory pressure Application of positive pressure to exhalation during mechanical ventilation; used to help prevent expiratory airway collapse and thus improve oxygenation.

Pressure gradient Difference in pressure between two regions.

PR interval Time between atrial depolarization and ventricular depolarization, as recorded on the ECG. The PR interval begins at the onset of the P wave and lasts until the onset of the QRS complex. Ordinarily, atrial contraction occurs during the PR interval.

Prosthetic valve Artificial heart valve replacing a nonfunctional valve. Ball-in-cage, bileaflet, porcine, and tilting-disk are four kinds of prosthetic heart valves.

Pulmonary artery Blood vessel leading from the right ventricle to the lungs.

Pulmonary artery dilation Expansion or stretching of the pulmonary artery beyond its normal dimensions.

Pulmonary circulation Blood pumped by the right ventricle into the pulmonary artery that circulates through the pulmonary capillary beds, where gas exchange occurs. The oxygenated blood is carried to the left atrium via the pulmonary veins.

Pulmonary edema Excessive accumulation of fluid within the lung.

Pulmonary hypertension Increased blood pressure within the pulmonary circulation.

Pulmonary vein One of four veins that return oxygenated blood from the lungs to the heart's left atrium.

Pulmonic ejection sound High-pitched click commonly created by the opening of a stenotic pulmonic valve. It is the only right-sided heart sound whose intensity increases during expiration and decreases during inspiration.

Pulmonic regurgitation Abnormal condition of backward blood flow across the pulmonic valve during diastole.

Pulmonic stenosis Narrowing or constriction of the pulmonic valve or the region just above or below the valve.

Pulmonic valvular stenosis Narrowing or constriction of the pulmonic valve.

Purkinje fibers Modified cardiac muscle cells found beneath the endocardium. They are part of the heart's electrical impulse-conducting system.

P wave Part of the ECG tracing that represents atrial depolarization.

QRS complex ECG waves representing the spread of electrical impulses from the bundle branches to the ventricular muscle. The QRS complex corresponds to ventricular depolarization.

Quality Heart sound characteristic determined by a combination of the sound's frequencies; one of six characteristics used to describe heart sounds. A sound's quality may be described as sharp, dull, booming, snapping, blowing, harsh, or musical.

Radiation Spread of a heart sound beyond its area of origin.

Rale *See* Crackle.

RBBB *abbr* Right bundle-branch block.

Regurgitant murmur Sound created by turbulent backward blood flow through an incompetent heart valve.

Repolarization Movement of calcium ions into the cell and potassium ions out of the cell, followed by the extrusion of sodium and calcium ions from the cell and the restoration of potassium ions into the cell by the sodium-potassium pump.

Resistance Force that hinders motion; hindrance or impedance.

Resonance Sound quality produced by percussing structures or cavities that radiate sound vibrations and energy.

Respiratory cycle One complete cycle of inspiration and expiration.

Rheumatic heart disease Valvular abnormalities that are an aftereffect of rheumatic fever; most commonly, mitral, tricuspid, or aortic stenosis and insufficiency.

Rhonchus *See* Low-pitched wheeze; plural form: Rhonchi.

Right bundle-branch block Interrupted conduction through the fiber that activates the right ventricle, resulting in a prolonged QRS complex.

Right mainstem bronchus One of two main branches extending from the trachea that supplies air to the right lung. It leaves the trachea at a less-acute angle than the left mainstem bronchus and is a more likely landing site for aspirated foreign bodies.

Right-sided heart failure Inability of the right ventricle to continue functioning properly.

Right-sided S_3 Heart sound caused by a noncompliant right ventricle and high right atrial pressure. It's heard in patients with cor pulmonale, pulmonary embolism, right-sided heart failure secondary to mitral stenosis with left-sided heart failure or pulmonary hypertension, and severe tricuspid insufficiency.

Right-sided S_4 S_4 generated in the right ventricle; commonly heard in conditions that create pressures greater than 100 mm Hg in that ventricle. It may accompany such conditions as pulmonic stenosis, pulmonary hypertension, or sudden tricuspid regurgitation.

Right ventricular outflow obstruction Stenosis, embolus, or other blockage preventing flow of deoxygenated blood from the right ventricle into the pulmonary artery.

S_1 *abbr* First heart sound; produced by closure of the mitral (M_1) and tricuspid (T_1) valves.

S_2 *abbr* Second heart sound; produced by closure of the aortic (A_2) and pulmonic (P_2) valves.

S_3 *abbr* Third heart sound; created by vibrations caused by the rapid, passive filling of the ventricles. S_3 is abnormal in adults older than age 20. The cadence is similar to the word *Tennessee*.

S_4 *abbr* Fourth heart sound; generated by stretching and filling of a ventricle during late diastole; associated with atrial contraction. The cadence is similar to the word *Kentucky*.

SA *abbr* Sinoatrial.

Sarcoidosis Granulomatous disease of unknown origin that may cause pulmonary fibrosis.

Scapula Triangular flat bone that makes up part of the shoulder girdle; also known as the *shoulder blade.*

Segmental bronchus Airway that branches from a lobar bronchus and conducts air to a lung segment.

SEM *abbr* Systolic ejection murmur (also referred to as SM).

Semilunar valves Crescent-shaped heart valves between the ventricles and the pulmonary artery and aorta.

Serous Having a watery consistency.

Shunting Abnormal communication between the high-pressure arterial system and the low-pressure venous system.

Silent chest Absence of breath sounds during auscultation; usually associated with severe bronchospasm and insufficient airflow to produce sounds.

Sinoatrial node Small mass of tissue at the junction of the superior vena cava and the right atrium that triggers the electrical impulses that begin the cardiac cycle.

SM *abbr* Systolic murmur (also referred to as SEM).

Snap Short, sharp heart sound associated with sudden closing or opening of a heart valve.

Status asthmaticus Severe asthma that's resistant to treatment; characterized by respiratory insufficiency or failure, wheezing, and severe dyspnea.

Stenosis Narrowing or constriction of a passage, specifically a heart valve or region around the outflow tract of any of the heart's chambers.

Sternal border Auscultatory area along and to either side of the sternum.

Stethoscope Instrument used in auscultation; usually consists of a diaphragm and a bell, which are connected to one or two tubes leading to a binaural headpiece and earpieces.

Stridor Noisy, high-pitched sound that can usually be heard at a distance from the patient; caused by laryngeal spasm and mucosal swelling, which contract the vocal cords and narrow the airway.

Stroke volume Amount of blood pumped during each ventricular contraction.

Subcutaneous emphysema Presence of air or gas in the tissues beneath the skin. It results in crackling or crepitus when touched.

Subvalvular aortic stenosis murmur Murmur caused by a left ventricular outflow obstruction below the aortic valve.

Subvalvular pulmonic stenosis Narrowing or constriction of the region below the pulmonic valve.

Subvalvular pulmonic stenosis murmur Murmur caused by a right ventricular outflow obstruction below the pulmonic valve.

Summation gallop S_4 that occurs simultaneously with S_3, thus sounding louder.

Superior vena cava Major vein that drains blood from the upper half of the body, beginning below the right costal cartilage and continuing to the right atrium.

Supine Lying on the back, face up.

Supravalvular aortic stenosis murmur Murmur caused by left ventricular outflow obstruction above the aortic valve.

Supravalvular pulmonic stenosis Narrowing or constriction of the region above the pulmonic valve.

Supravalvular pulmonic stenosis murmur Murmur caused by a right ventricular outflow obstruction above the pulmonic valve.

Surfactant Active surface agent that acts to decrease surface tension, thereby preventing alveolar collapse; believed to be secreted by type 2 alveolar cells.

Sympathetic nervous system Part of the nervous system that supplies innervation to the visceral and musculoskeletal system. Its primary hormone transmitter is norepinephrine. Sympathetic innervation to the airways facilitates smooth-muscle relaxation, thereby causing bronchial dilation.

Systemic hypertension Condition in which the patient has a higher-than-normal overall blood pressure.

Systole Ventricular contraction that ejects blood into the arterial system. The left ventricle ejects into the aorta, and the right ventricle ejects into the pulmonary artery.

Systolic crescendo murmur Heart sound that begins as a faint sound and then rises in volume; occurs during the ventricular contraction phase of the cardiac cycle.

Systolic ejection murmur Murmur heard in systole that's caused by turbulent blood flow as the right or left ventricle ejects blood; it can be an innocent murmur.

Systolic regurgitation murmur Sound created by turbulent backward blood flow from either the mitral or tricuspid valve during the ventricular contraction phase of the cardiac cycle.

T_1 *abbr* Tricuspid component of S_1.

Terminal respiratory bronchioles Final part of the conducting airways. They mark the beginning of the respiratory zone.

Thoracic cavity Space within the rib cage that begins at the clavicle and ends at the diaphragm.

Thorax Bony structure that encloses the thoracic cavity, protecting the heart, lungs, and great vessels.

3LSB *abbr* Third intercostal space, left sternal border (Erb's point).

Thrill Abnormal tremor felt on palpation that accompanies some vascular or cardiac murmur.

Tilting-disk valve Prosthetic heart valve characterized by a disk that tilts with ventricular pressure to open and close the valvular opening.

Tracheal breath sound Loud, tubular (hollow) breath sound auscultated over the trachea that's audible during inspiration and expiration.

Tracheobronchial tree Portion of the airway that begins at the larynx and ends at the terminal bronchioles; also known as the *lower airway.*

Tricuspid insufficiency Tricuspid valve incompetence during systole, which allows turbulent backward, or regurgitant, blood flow into the right atrium.

Tricuspid stenosis Narrowing or constriction of the tricuspid valve.

Tricuspid valve Heart valve between the right atrium and right ventricle.

Tubular breath sound Loud, hollow sound characteristically heard over the trachea and mainstem bronchi.

Turbinate *See* Conchae.

Turbulence Disturbed or irregular airflow; can be caused by rapid flow rates or variations in air pressures and velocities.

T wave Part of the ECG tracing that represents ventricular repolarization.

2LSB *abbr* Second intercostal space, left sternal border (pulmonic area)

2RSB *abbr* Second intercostal space, right sternal border (aortic area)

Type I alveolar cell One of three cell types that make up the alveoli; covers about 95% of the alveolar surface area.

Type II alveolar cell One of three cell types that line the alveoli; source of pulmonary surfactant.

Valsalva's maneuver Attempt to exhale forcibly with the glottis closed, resulting in increased intrathoracic pressure, reduced heart rate and venous return, and increased venous pressure.

Ventilation Movement of air in and out of the lungs.

Ventricular ejection Ventricular contraction that sends blood into the arterial system; called *systole* in the cardiac cycle.

Ventricular filling Ventricular relaxation and expansion that allows blood to enter; called *diastole* in the cardiac cycle.

Ventricular outflow obstruction Blockage in the valve or vessel that carries blood out of one of the heart's ventricles.

Ventricular septal defect Opening in the septum between the ventricles; usually represents a congenital abnormality.

Ventricular tachycardia Rapid heart rate originating from one or more ectopic foci in the ventricle.

Vesicular sounds Term commonly used to describe normal breath sounds auscultated over most of the chest wall.

Vocal cord One of two membranous structures in the larynx responsible for phonation.

Vortex Circular airflow caused by the shearing force of high-velocity airstreams alongside slower airstreams. Vortices within the airways are precipitated by airway branching that causes airflow to change direction abruptly.

Wheeze Continuous, high-pitched sound that has a musical quality; results from bronchospasm.

Whispered pectoriloquy High-frequency whispered voice sound auscultated over consolidated or atelectatic areas.

Index

i refers to an illustration; t refers to a table.

---

i refers to an illustration; t refers to a table.

---

i refers to an illustration; t refers to a table.

i refers to an illustration; t refers to a table.

---

i refers to an illustration; t refers to a table.

i refers to an illustration; t refers to a table.

i refers to an illustration; t refers to a table.

i refers to an illustration; t refers to a table.